Brittle Materials in Mechanical Extremes

Brittle Materials in Mechanical Extremes

Editor

Giovanni Bruno

MDPI • Basel • Beijing • Wuhan • Barcelona • Belgrade • Manchester • Tokyo • Cluj • Tianjin

Editor
Giovanni Bruno
Bundesanstalt für Materialforschung und –prüfung (BAM)
Germany

Editorial Office
MDPI
St. Alban-Anlage 66
4052 Basel, Switzerland

This is a reprint of articles from the Special Issue published online in the open access journal *Materials* (ISSN 1996-1944) (available at: https://www.mdpi.com/journal/materials/special_issues/brittle_materials_mechanical_extremes).

For citation purposes, cite each article independently as indicated on the article page online and as indicated below:

LastName, A.A.; LastName, B.B.; LastName, C.C. Article Title. *Journal Name* **Year**, *Volume Number*, Page Range.

ISBN 978-3-03943-927-0 (Hbk)
ISBN 978-3-03943-928-7 (PDF)

© 2020 by the authors. Articles in this book are Open Access and distributed under the Creative Commons Attribution (CC BY) license, which allows users to download, copy and build upon published articles, as long as the author and publisher are properly credited, which ensures maximum dissemination and a wider impact of our publications.

The book as a whole is distributed by MDPI under the terms and conditions of the Creative Commons license CC BY-NC-ND.

Contents

About the Editor . . . . **vii**

Giovanni Bruno
Brittle Materials in Mechanical Extremes
Reprinted from: *Materials* 2020, 13, 4610, doi:10.3390/ma13204610 1

Liang Li, Wenli Liu, Jun Wu, Wenjie Wu and Meng Wu
Experimental Investigation on the Quasi-Static Tensile Capacity of Engineered Cementitious Composites Reinforced with Steel Grid and Fibers
Reprinted from: *Materials* 2019, 12, 2666, doi:10.3390/ma12172666 5

Yangsheng Ye, Gang Xu, Liangwei Lou, Xianhua Chen, Degou Cai and Yuefeng Shi
Evolution of Rheological Behaviors of Styrene-Butadiene-Styrene/Crumb Rubber Composite Modified Bitumen after Different Long-Term Aging Processes
Reprinted from: *Materials* 2019, 12, 2345, doi:10.3390/ma12152345 27

Angelo Masi, Andrea Digrisolo and Giuseppe Santarsiero
Analysis of a Large Database of Concrete Core Tests with Emphasis on Within-Structure Variability
Reprinted from: *Materials* 2019, 12, 1985, doi:10.3390/ma12121985 45

Francesco Baino
Quantifying the Adhesion of Silicate Glass–Ceramic Coatings onto Alumina for Biomedical Applications
Reprinted from: *Materials* 2019, 12, 1754, doi:10.3390/ma12111754 61

Matti Isakov, Janin Lange, Sebastian Kilchert and Michael May
In-Situ Damage Evaluation of Pure Ice under High Rate Compressive Loading
Reprinted from: *Materials* 2019, 12, 1236, doi:10.3390/ma12081236 73

René Laquai, Fanny Gouraud, Bernd Randolf Müller, Marc Huger, Thierry Chotard, Guy Antou and Giovanni Bruno
Evolution of Thermal Microcracking in Refractory ZrO_2-SiO_2 after Application of External Loads at High Temperatures
Reprinted from: *Materials* 2019, 12, 1017, doi:10.3390/ma12071017 89

Belynda Benane, Sylvain Meille, Geneviève Foray, Bernard Yrieix and Christian Olagnon
Instrumented Indentation of Super-Insulating Silica Compacts
Reprinted from: *Materials* 2019, 12, 830, doi:10.3390/ma12050830 105

Vladimir Buljak, Tyler Oesch and Giovanni Bruno
Simulating Fiber-Reinforced Concrete Mechanical Performance Using CT-Based Fiber Orientation Data
Reprinted from: *Materials* 2019, 12, 717, doi:10.3390/ma12050717 119

Tai Thanh Tran and Hyug-Moon Kwon
Influence of Activator Na_2O Concentration on Residual Strengths of Alkali-Activated Slag Mortar upon Exposure to Elevated Temperatures
Reprinted from: *Materials* 2018, 11, 1296, doi:10.3390/ma11081296 135

He Zhu, Qingbin Li, Yu Hu and Rui Ma
Double Feedback Control Method for Determining Early-Age Restrained Creep of Concrete Using a Temperature Stress Testing Machine
Reprinted from: *Materials* **2018**, *11*, 1079, doi:10.3390/ma11071079 . . . . **155**

About the Editor

Giovanni Bruno, Born 1966 in Potenza, Italy, he studied Nuclear Engineering (1989) and then Physics (1998) at the University of Bologna, Italy. He received his Ph.D. in Materials Science at the University of Ancona, Italy, in 1997, working on residual stress analysis in welds of several metallic alloys. He worked as a Post-Doctorate in the UK, in Germany, and in France, dealing with residual stress analysis and mechanical properties of metals and ceramics. He then moved to Corning Incorporated, where he worked first as the head of the Physical, Mechanical and Structural Characterization Group in France and then as a project leader in the United States. He has been Head of Division 8.5 Micro NDT at BAM (Berlin, D) and a professor at the University of Potsdam (D) as of 2012. His research interests are focused on the use of microstructures as determined by computed tomography and on the determination of residual stress, mechanical properties and damage mechanisms of composites, metallic alloys (in particular, additively manufactured) and ceramics.

Editorial

Brittle Materials in Mechanical Extremes

Giovanni Bruno

BAM, Bundesanstalt für Materialforschung und prüfung, Unter den Eichen, 87, 12205 Berlin, Germany; Giovanni.bruno@bam.de; Tel.: +49-30-8104-1850

Received: 9 October 2020; Accepted: 14 October 2020; Published: 16 October 2020

Abstract: The goal of the Special Issue "Brittle Materials in Mechanical Extremes" was to spark a discussion of the analogies and the differences between different brittle materials, such as, for instance, ceramics and concrete. Indeed, the contributions to the Issue spanned from construction materials (asphalt and concrete) to structural ceramics, reaching as far as ice. The data shown in the issue were obtained by advanced microstructural techniques (microscopy, 3D imaging, etc.) and linked to mechanical properties (and their changes as a function of aging, composition, etc.). The description of the mechanical behavior of brittle materials under operational loads, for instance, concrete and ceramics under very high temperatures, offered an unconventional viewpoint on the behavior of brittle materials. This is not at all exhaustive, but a way to pave the road for intriguing and enriching comparisons.

Keywords: ceramics; concrete; asphalt; mechanical properties; microstructure; microcracking; strength

---

As a premise, it must be said that brittle materials are such an enormous category of materials that it would be impossible to disclose and even comment on their mechanical properties in a short editorial. The following text is therefore limited to the contents of the Special Issue "Brittle Materials in Mechanical Extremes", and to some of the works that the editor has conducted, leading to the initiation of the Special Issue.

Brittle materials include a wide range of material classes: From polymers to metals, through to classic glass, ceramics, and composites. They all share a supposed linear elastic behavior but are often found to display non-linear stress–strain relationships or high temperature dilation (or other properties such as thermal conductivity). In this Special Issue, contributions describing and explaining this intriguing behavior, whether due to microcracking, interaction among constituent phases, or micro-structural features (such as pores), were collated. Advanced characterization techniques, challenging numerical and analytical models, unconventional experiments, and/or the analysis of existing field data were reported and should spark the debate about the origin of the mechanical behavior of brittle materials under various and unconventional mechanical, weathering, and thermal loads.

The Special Issue allows (through the large amount of experimental data provided by the contributing authors) a discussion of the analogies and the differences between different materials, such as, e.g., ceramics and concrete. Data have been corroborated by advanced microstructural studies (microscopy, 3D imaging, etc.), sometimes leading to the identification of the microstructure–property relationships through appropriate models.

A few works have demonstrated the peculiar behavior of brittle materials, such as concrete (see the book of Torrenti et al. [1]), composite materials (see e.g., [2]), and ceramics (see e.g., [3,4]), especially if they possess special features such as pores or hard phases, or undergo microcracking. Microcracks, together with porosity, impart to such materials an exceedingly large strain tolerance (strain at failure), where the specimen is able to carry some (small) load even when it is fully fractured (thereby displaying residual strength [5]), excellent corrosion resistance [6], or extremely high thermal shock resistance [7]. In this sense, such materials behave similarly to metamaterials, where the meso-structure (at scales above the grain size) is as important as the material composition of the micro-structure (at scale below the grain size). Simply thinking of microcracked glass or ceramics, possessing negative macroscopic thermal expansion [8], exemplifies such astonishing behavior.

Several techniques have been used in the literature (and were used in this Special Issue) to characterize this behavior: instrumented indentation (see, for instance, the book of Buljak [9], or the contribution [10] to this Special Issue) is particularly suitable to monitor the variation of the Young's modulus as a function of the load, thereby allowing insights into "mechanical microcracking" (see [11] for its definition). Moreover, indentation allows for the use of inverse methods to extract the (non-linear) constitutive behavior of materials [12]. If the results obtained by the application of inverse methods are verified by independent experiments (as done for example in Reference [13]), the fitting parameters acquire the meaning of materials' properties. An example of this approach is also given in the Special Issue [14]. In general, modeling is now a requirement for good experimental data to be fully exploited. Coupling even simple models to data on the mechanical behavior extends the validity of such data to a more universal level. This is what was practiced in this Special Issue, whether with creep models for concrete or damage models for asphalt [15], thereby setting a standard for good scientific practice.

From an experimental point of view, this Special Issue demonstrated that inventive is needed to solve the problem of the characterization of the relevant quantities in brittle in materials. Special set-ups [16] or unconventional test methods [17] were utilized to disclose the damage behavior of such materials. In order to characterize the microstructure, the use of 3D imaging techniques (usually X-ray based) is becoming increasingly standardized in materials science to disclose meso-structures (pores, cracks, inhomogeneities), while the microstructural characterization is still classically based on optical microscopy (OM) and scanning electron microscopy (SEM) pictures. Composite materials, porous ceramics and concrete all carry meso-structural features (again, with sizes above the crystal grain) that can be well imaged by laboratory computed tomography. Apart from carrying an augmented information content, 3D data such as those from X-ray computed tomography, optical tomography or 3D infra-red thermography, allow the use of finite element simulations based on experimentally determined microstructures, or the extraction of quantitative data (e.g., fiber orientation distribution in concrete [18]) at the scale of a representative volume element (RVE). On top of that, in this Special Issue, the strength of a relatively novel technique, X-ray refraction radiography (see [19] for an introduction to it) was successfully used to quantify damage in thermally cycled refractory ceramics [20].

Finally, the description of the mechanical behavior of brittle materials under operational (sometimes unconventional) loads, such as mechanical and temperature cycling, electric fields, corrosion environments, while already underway in the literature, represents the natural extension of this Special Issue.

Funding: This research received no external funding.

Conflicts of Interest: The authors declare no conflict of interest.

References

1. Torrenti, J.M.; Pijaudier-Cabot, G.; Reynouard, J.M. *Mechanical Behavior of Concrete*; Wiley & Sons Inc.: Hoboken, NJ, USA, 2010. [CrossRef]
2. Evsevleev, S.; Cabeza, S.; Mishurova, T.; Garcés, G.; Sevostianov, I.; Requena, G.; Boin, M.; Hofmann, M.; Bruno, G. Stress-induced damage evolution in cast AlSi12CuMgNi alloy with one and two ceramic reinforcements. Part II: Effect of reinforcement orientation. *J. Mater. Sci.* **2020**, *55*, 1049–1068. [CrossRef]
3. Liens, A.; Reveron, H.; Douillard, T.; Blanchard, N.; Lughi, V.; Sergo, V.; Laquai, R.; Müller, R.; Bruno, B.G.; Schomer, S.; et al. Phase transformation induces plasticity with negligible damage in ceria-stabilized zirconia-based ceramics. *Acta Mater.* **2020**, *183*, 261–273. [CrossRef]

4. Bruno, G.; Efremov, A.M.; Levandovskiy, A.N.; Pozdnyakova, I.; Hughes, D.J.; Clausen, B. Thermal and Mechanical Response of Industrial Porous Ceramics. *Mat. Sci. Forum* **2010**, *652*, 191–196. [CrossRef]
5. Babelot, C.; Guignard, A.; Huger, M.; Gault, C.; Chotard, T.; Ota, T.; Adachi, N. Preparation and thermomechanical characterisation of aluminum titanate flexible ceramics. *J. Mater. Sci.* **2011**, *46*, 1211–1219. [CrossRef]
6. Trümer, A.; Ludwig, H.-M. Sulphate and ASR Resistance of Concrete Made with Calcined Clay Blended Cements. In Proceedings of the 1st International Conference on Calcined Clays for Sustainable Concrete, Zürich, Switzerland, 23–25 June 2015.
7. Hasselman, D.P.H. Unified theory of thermal shock fracture initiation and crack propagation in brittle ceramics. *J. Am. Ceram. Soc.* **1969**, *52*, 600–604. [CrossRef]
8. Holand, W.; Beall, G.H. *Glass Ceramic Technology*, 2nd ed.; Wiley & Sons Inc.: Hoboken, NJ, USA, 2012.
9. Buljak, V. *Inverse Analyses with Model Reduction: Proper Orthogonal Decomposition in Structural Mechanics*; Springer: Berlin, Germany, 2011.
10. Benane, B.; Meille, S.; Foray, G.; Yrieix, B.; Olagnon, C. Instrumented Indentation of Super-Insulating Silica Compacts. *Materials* **2019**, *12*, 830. [CrossRef] [PubMed]
11. Bruno, G.; Efremov, A.M.; An, C.; Nickerson, S. Not All Microcracks Are Born Equal: Thermal vs Mechanical Microcracking In Porous Ceramics. In *Advances in Bioceramics and Porous Ceramics IV: Ceramic Engineering and Science Proceedings*; The American Ceramic Society: Westerville, OH, USA, 2011; Volume 32.
12. Neto, F.D.M.; Neto, A.J.S. *An Introduction to Inverse Problems with Applications*; Springer: Heidelberg, Germany, 2013.
13. Buljak, V.; Bruno, G. Numerical modeling of thermally induced microcracking in porous ceramics: An approach using cohesive elements. *J. Eur. Ceramic Soc.* **2018**, *38*, 4099–4108. [CrossRef]
14. Buljak, V.; Oesch, T.; Bruno, G. Simulating Fiber-Reinforced Concrete Mechanical Performance Using CT-Based Fiber Orientation Data. *Materials* **2019**, *12*, 717. [CrossRef] [PubMed]
15. Ye, Y.; Xu, G.; Lou, L.; Chen, X.; Cai, D.; Shi, Y. Evolution of Rheological Behaviors of Styrene-Butadiene-Styrene/Crumb Rubber Composite Modified Bitumen after Different Long-Term Aging Processes. *Materials* **2019**, *12*, 2345. [CrossRef] [PubMed]
16. Zhu, H.; Li, Q.; Hu, Y.; Ma, R. Double Feedback Control Method for Determining Early-Age Restrained Creep of Concrete Using a Temperature Stress Testing Machine. *Materials* **2019**, *12*, 1754. [CrossRef] [PubMed]
17. Isakov, M.; Lange, J.; Kilchert, S.; May, M. In-Situ Damage Evaluation of Pure Ice under High Rate Compressive Loading. *Materials* **2019**, *12*, 1236. [CrossRef] [PubMed]
18. Mishurova, T.; Rachmatulin, N.; Fontana, P.; Oesch, T.; Bruno, G.; Radi, E.; Sevostianov, I. Evaluation of the probability density of inhomogeneous fiber orientations by computed tomography and its application to the calculation of the effective properties of a fiber-reinforced composite. *Int. J. Eng. Sci.* **2018**, *122*, 14–29. [CrossRef]
19. Kupsch, A.; Müller, B.R.; Lange, A.; Bruno, G. Microstructure characterisation of ceramics via 2D and 3D X-ray refraction techniques. *J. Eur. Ceram. Soc.* **2017**, *37*, 1879–1889. [CrossRef]
20. Laquai, R.; Gouraud, F.; Müller, B.R.; Huger, M.; Chotard, T.; Antou, G.; Bruno, G. Evolution of Thermal Microcracking in Refractory ZrO2-SiO2 after Application of External Loads at High Temperatures. *Materials* **2019**, *12*, 1017. [CrossRef] [PubMed]

Publisher's Note: MDPI stays neutral with regard to jurisdictional claims in published maps and institutional affiliations.

© 2020 by the author. Licensee MDPI, Basel, Switzerland. This article is an open access article distributed under the terms and conditions of the Creative Commons Attribution (CC BY) license (http://creativecommons.org/licenses/by/4.0/).

Article

Experimental Investigation on the Quasi-Static Tensile Capacity of Engineered Cementitious Composites Reinforced with Steel Grid and Fibers

Liang Li [1,*], Wenli Liu [1], Jun Wu [2,*], Wenjie Wu [1] and Meng Wu [2]

1. Key Laboratory of Urban Security and Disaster Engineering, Beijing University of Technology, Ministry of Education, Beijing 100124, China
2. School of Urban Railway Transportation, Shanghai University of Engineering Science, Shanghai 201620, China
* Correspondence: liliang@bjut.edu.cn (L.L.); cvewujun@sues.edu.cn (J.W.); Tel.: +86-10-6739-2430 (L.L.)

Received: 4 July 2019; Accepted: 14 August 2019; Published: 21 August 2019

Abstract: An engineered cementitious composite (ECC) was reinforced with a steel grid and fibers to improve its tensile strength and ductility. A series of tensile tests have been carried out to investigate the quasi-static tensile capacity of the reinforced ECC. The quasi-static tensile capacities of reinforced ECCs with different numbers of steel-grid layers, types of fibers (Polyvinyl alcohol (PVA) fiber, KEVLAR fiber, and polyethylene (PE) fiber), and volume fractions of fibers have been tested and compared. It is indicated by the test results that: (1) On the whole, the steel grid-PVA fiber and steel grid-KEVLAR fiber reinforced ECCs have high tensile strength and considerable energy dissipation performance, while the steel grid-PE fiber reinforced ECC exhibits excellent ductility. (2) The ultimate tensile strength of the reinforced ECC can be improved by the addition of steel grids. The maximal peak tensile stress increase is about 50–95% or 140–190% by adding one layer or two layers of steel grid, respectively. (3) The ultimate tensile strength of the reinforced ECC can be enhanced with the increase of fiber volume fraction. For a certain kind of fiber, a volume fraction between 1.5% and 2% grants the reinforced ECC the best tensile strength. Near the ultimate loading point, the reinforced ECC exhibits strain hardening behavior, and its peak tensile stress increases considerably. The energy dissipation performance of the reinforced ECC can also be remarkably enhanced by such an increase in fiber volume fraction. (4) The ductility of the steel grid-PVA fiber reinforced ECC can be improved by the addition of steel grids and the increase of fiber volume fraction. The ductility of the steel grid-KEVLAR fiber reinforced ECC can be improved by the addition of steel grids alone. The ductility and energy dissipation performance of the steel grid-PE fiber reinforced ECC can be improved with the increase of fiber volume fraction alone. A mechanical model for the quasi-static initial and ultimate tensile strength of the steel grid-fiber reinforced ECC is proposed. The model is validated by the test data from the quasi-static tension experiments on the steel grid-PE fiber reinforced ECC.

Keywords: engineered cementitious composites; steel grid; fiber; tensile capacity; energy dissipation

1. Introduction

Engineered cementitious composites (ECCs) have been used widely in the civil engineering and transportation applications, such as airport runways. They have a higher tensile strength and ductility compared to normal concrete. They also exhibit high energy dissipation performance due to strain hardening and multiple-cracking behavior [1]. During the past decades, various types of ECCs with different ingredients have been developed. The mechanical properties of these ECCs have also been the subject of intense research during the past decades.

Various types of fibers, such as carbon fibers, steel fibers, and polymer fibers have been added to the ECCs to improve their tensile capacity and energy dissipation performance. Hybrid fibers, namely, the combination of different types of fibers, have also been adopted to gain the composite effect. Tran and Kim [2] investigated the direct tensile stress versus strain response of high-performance fiber-reinforced cementitious composites (HPFRCCs) at high strain rates between $10s^{-1}$ and $40s^{-1}$. Twisted and hooked steel fibers were used in the HPFRCCs. Arboleda et al. [3] studied the tensile behavior of fabric-reinforced cementitious matrix composites. Different fabrics, including polyparaphenylene benzobisoxazole (PBO), carbon, glass, and carbon and glass with a special protective coating were used for the investigation. Kim et al. [4] conducted direct tensile and shear transfer tests of amorphous micro steel (AMS) fiber-reinforced cementitious composites. Ali et al. [5,6] investigated the behavior under impact loading of a hybrid fiber-reinforced ECC incorporating short, randomly dispersed shape memory alloy (HECC-SMAF) and PVA fibers by the drop weight impact test and numerical simulation. Yu et al. [7,8] developed ultra-high ductile cementitious composites (UHDCCs) with the polyethylene (PE) fibers. It was reported that the tested UHDCCs exhibited an average tensile strain of 8% at peak stress. The rate sensitivity of the UHDCCs was evaluated by direct tensile experiments under different strain rates. Curosu et al. [9] investigated the tensile behavior of high-strength strain-hardening cement-based composites (HS-SHCCs) made with four different types of dispersed high-performance polymer fibers.

Zhou et al. [10] investigated the mechanical properties of hybrid ECCs incorporating steel and polyethylene fibers. Zhang et al. [11] studied the mechanical properties and carbonation durability of ECCs reinforced by polypropylene and hydrophilic polyvinyl alcohol fibers. Kim et al. [12] investigated the hybrid effect of twisted steel and polyethylene fibers on the tensile performance of ECCs. Zhu et al. [13] conducted uniaxial tensile tests to investigate the stress-strain behavior of carbon-fiber grid-reinforced ECCs. Al-Gemeel et al. [14] and Li et al. [15] investigated the tensile behavior of a basalt textile grid reinforced ECC. Sun et al. [16] conducted a series of tests to study the mechanical behavior of the ECCs reinforced with polyvinyl alcohol (PVA) fibers.

Some research has been focused on the factors that impact the tensile performance of ECCs. Wang et al. [17] studied the tensile performance of polyvinyl alcohol (PVA)-steel hybrid fiber reinforced ECCs, focusing on the impacts of steel-fiber content and water-to-binder ratio of the matrix. Abrishambaf et al. [18] investigated the influence of fiber orientation on the tensile behavior of ultra-high-performance fiber-reinforced cementitious composites. Wu et al. [19] studied the effect of the morphological parameters of natural sand on the mechanical properties of ECCs. The mechanical behavior of ECC and fiber-reinforced ECC at high temperatures has also been investigated for solar emission, fire, gas explosion, and blast scenarios [20,21].

Even the tensile strength and ductility of normal ECC or fiber-reinforced ECC are not sufficient to resist strong blast and impact loads. The tensile strength and energy dissipation performance of ECC can be improved by adding a steel grid and fibers to the matrix. In the current study, the ECC is reinforced with a steel grid and fibers to improve its tensile strength and ductility. A series of experiments are carried out to investigate the quasi-static tensile strength of the reinforced ECC using a Z100 material tensile testing machine manufactured by the Zwick/Roell Group (Ulm, Germany). The quasi-static tensile strength and energy dissipation performance of the various reinforced ECCs are tested and compared. The test variations include the number of steel-grid layers (one layer, two layers), the type of fibers (polyvinyl alcohol (PVA) fiber, KEVLAR fiber, and polyethylene (PE) fiber), and the volume fraction of fibers (0%, 0.5%, 1%, 1.5%, 2%). A mechanical model for the quasi-static initial and ultimate tensile strength of the steel grid-fiber reinforced ECC is proposed. This model is validated by the test data from the quasi-static tension experiments on the steel grid-PE fiber reinforced ECC.

2. Experimental Program

2.1. Materials and Mixture Proportions

The raw materials used in the current study for the matrix of the ECC include P.O42.5 ordinary Portland cement, siliceous fly ash, water, and superplasticizer. Three types of fibers, (polyvinyl alcohol (PVA) fibers, KEVLAR fibers, and polyethylene (PE) fibers) and two different steel-grid configurations (one layer and two layer) were added to the matrix to improve its tensile strength and ductility. The physical and mechanical properties of P.O42.5 ordinary Portland cement are listed in Table 1. The term "initial setting time" refers to the time required for the cement slurry to begin losing plasticity, and the term "final setting time" refers to the time required for the cement slurry to completely lose its plasticity and to begin to exhibit considerable strength. The three types of fibers used in the experiments are shown in Figure 1, and their basic physical parameters are presented in Tables 2–4. The steel grids used in the experiments are shown in Figure 2, the one-layer steel grid is on the right side, and the two-layer one is on the left side. The diameter of steel grid wire is 0.88 mm, the dimensions of the grid holes are 12.83 mm × 12.76 mm.

The mass mixture proportions of cement, silica ash, water, and superplasticizer for the matrix of the ECC are listed in Table 5. The matrix is reinforced with one layer or two layers of steel grid, respectively. In addition, one of the three types of fibers mentioned above is added to the matrix of the ECC. The volume contents of fiber in the current experimental study are 0%, 0.5%, 1%, 1.5%, and 2.0%, respectively.

Table 1. Physical and mechanical properties of P.O42.5 ordinary Portland cement.

Specific Surface Area	Initial Setting Time	Final Setting Time	Compression Strength (3 Days)	Bending Strength (3 Days)
(m²/kg)	(min)	(min)	(MPa)	(MPa)
381	181	243	23.5	5.3

Table 2. Basic physical parameters of PVA fiber.

Diameter	Standard Length	Tensile Strength	Elongation Ratio	Elastic Modulus	Density
(μm)	(mm)	(MPa)		(GPa)	(g/cm³)
40	12	1560	6.5%	41	1.3

Table 3. Basic physical parameters of KEVLAR fiber.

Tensile Strength	Elastic Modulus	Elongation Ratio	Standard Length	Density
(MPa)	(GPa)		(mm)	(g/cm³)
2920	70.5	3.6%	12	1.44

Table 4. Basic physical parameters of PE fiber.

Tensile Strength	Elastic Modulus	Elongation Ratio	Standard Length	Density
(GPa)	(GPa)		(mm)	(g/cm³)
2.18	66	3.5%	12	0.97

Table 5. Mass mixture proportions for the matrix of ECC.

Cement	Siliceous Fly Ash	Water	Superplasticizer
1.0	0.11	0.3	0.013

Figure 1. Three types of fibers used in the current study. (**a**) PVA fiber; (**b**) KEVLAR fiber; (**c**) PE fiber.

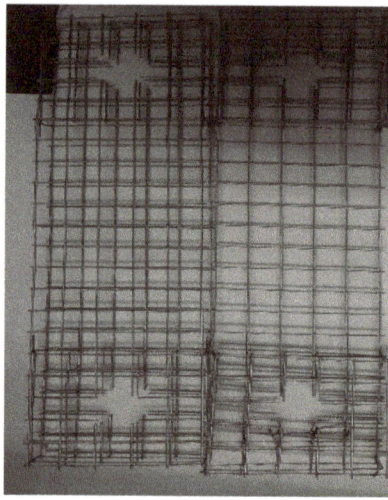

Figure 2. Double-layer (**left**) and single-layer (**right**) steel grids used in the current study.

2.2. Specimen Fabrication and Test Setup

A rectangular, thin-plate-shaped specimen is adopted in the current experimental study, and the dimensions of the specimen are 300 mm × 75 mm × 20 mm. A steel mold is used for the casting of the specimens. The specimen fabrication procedure can be described as follows:

(1). Spread lubricant on the inner surface of the mold for the convenience of the demolding process. Install the steel grid in the middle part of the mold.
(2). Weigh-up the material ingredients of the tested ECCs, including cement, siliceous fly ash, water, superplasticizer, and fibers according to the mixture proportion. When the KEVLAR fiber is used, a cleaning process should be carried out with alcohol to remove grease from the fiber surfaces.
(3). The cement and siliceous fly ash are first dry-mixed in the mixer for 3 min. The superplasticizer is mixed with the water, and they are then added into the dry mixture and mixed for a further 2 min to produce a consistent and uniform matrix. The fibers are then added into the matrix and mixed for an additional 3 min to make the fibers spread in the mixture until they reach a uniform state.
(4). The fresh ECC mixtures are cast into the steel molds. The molds are then placed on a shake-table to eject entrapped air and produce a denser matrix by the vibration of the table-board. The specimens are demolded after 24 h and then set in a curing room for 28 days, where the temperature is 20 ± 0.5 °C, and the relative humidity is 95 ± 5%.
(5). When the curing process is finished, thin steel pieces are adhered to the ends of the specimens with epoxy resin adhesive in order to provide extra reinforcement at the location where the specimens are connected to the tensile-loading machine. The specimens with the end reinforcement are shown in Figure 3.

Figure 3. Specimens with the end reinforcement.

The Z100 universal material testing machine manufactured by the Zwick/Roell Group (Ulm, Germany) is used for the tensile tests of the ECCs reinforced with steel grid and fibers. The displacement-controlled loading mode is adopted for the test. The tensile load is measured by the 100kN force transducer. A pair of automatic extensometers is used to measure the strain of the specimen. The measurement range of each extensometer is 100 mm.

The tensile loading rate is 0.1mm/min, and the corresponding strain rate is 1×10^{-5} s^{-1}. The tensile stress and strain of the specimen can be computed as Equations (1) and (2):

$$\sigma = F/bh \qquad (1)$$

$$\varepsilon = L/L_0 \qquad (2)$$

where σ and ε are the tensile stress and strain of the specimen, F is the measured tensile force, b and h are the width and thickness of the specimen, respectively. In the current test study, $b = 75$ mm, $h = 20$ mm. L is the measured tensile deformation value of the specimen, and L_0 is the standard distance between two extensometers, in the current test $L_0 = 100$ mm.

3. Test Results and Discussion

3.1. Quasi-Static Tensile Test Results of the Steel Grid-PVA Fiber Reinforced ECC

The quasi-static tensile test results of the ECC reinforced with steel grids and PVA fibers are presented in Table 6. Because ECC exhibits considerable strength and ductility even after reaching its peak tensile stress, during loading, the failure of ECC is considered to occur when the tensile stress has descended to 80% of its peak value. The strain corresponding to the 80% of the peak stress in the descending segment of the stress-strain curve is defined as ultimate strain. The energy dissipation can be computed from the area enveloped by the stress–strain curve. In the specimen notation, "M" stands for the matrix, "S" stands for the steel grid, "A" stands for the PVA fiber, and the two numbers stand for the volume content of the PVA fibers and the number of steel-grid layers, respectively. For example, "A0.5S2" stands for the ECC specimen with a volume content of the PVA fibers of 0.5% and two layers of steel grid.

A comparison of the energy dissipation of different types of steel grid-PVA fiber reinforced specimens is shown in Figure 4. It is illustrated that the energy dissipation performance of the ECC can be improved remarkably by the addition of a steel grid. For the ECC matrix specimens, only a single, main crack is observed and the specimen splits along the crack when the failure occurs. For the steel grid-PVA fiber reinforced ECC specimens, however, a multiple-cracking phenomenon can be observed during the tensile loading. Some post-failure specimens with multiple cracks are shown in Figure 5. After the appearance of the first crack, the tension stress in the cracking region is mainly borne by the steel grid, which restrains the further development of a single, critical crack.

Table 6. Quasi-static tensile test results of the steel grid-PVA fiber reinforced ECC.

Specimen Type	Initial Cracking Stress (MPa)	Peak Stress (MPa)	Ultimate Strain (%)	Energy Dissipation (J/m^2)
M	0.98	0.98	0.01	1578
M-S1	0.86	1.29	0.72	11,896
M-S2	1.77	2.33	0.32	265,793
A0.5	1.39	1.39	0.05	88,070
A0.5S1	1.56	2.13	0.58	749,070
A0.5S2	3.34	3.34	0.60	1,072,785
A1	1.77	1.92	0.35	188,565
A1S1	2.05	2.58	0.28	557,326
A1S2	1.58	2.99	0.50	1,121,904
A1.5	1.78	2.00	0.23	201,600
A1.5S1	3.89	3.90	1.3	1,248,030
A1.5S2	5.14	5.14	0.64	880,790
A2	2.63	2.80	0.39	568,008
A2S1	3.93	3.93	0.67	796,900
A2S2	5.73	5.73	1.08	1,494,499

A comparison of the stress-strain curves of the specimens with different numbers of steel-grid layers is shown in Figure 6. A comparison for the specimens with different fiber volume fractions is shown in Figure 7. These results indicate that the tensile strength of the PVA-ECC can be enhanced by the addition of steel grids. For the fiber volume fraction of 1.5%, the maximal peak tensile stress increase of about 95% or 160% compared to the matrix specimen by adding one layer or two layers of steel grid can be obtained, respectively. In addition, the ductility of the PVA-ECC can also be improved to some extent by the addition of steel grids. The ultimate strain increases by about 0.6% or 0.8% by adding one layer or two layers of steel grid, respectively. This means that the ultimate strain of the steel-grid specimens is increased by 0.6% or 0.8% absolute strain compared to the matrix specimens. After the peak stress, the steel grids begin to separate from the matrix, and this leads to a deformation inconsistency between the steel grids and the matrix. The peak tensile stress of the steel grid-PVA fiber reinforced ECC can be enhanced with the increase of the volume fraction of PVA fibers. For the volume fractions of 0.5%, 1%, 1.5%, and 2%, the peak tensile stress increases by about 45%, 80%, 85%, and 170% compared to the matrix specimen, respectively. The ductility can also be improved with the increase of the volume fraction of PVA fibers.

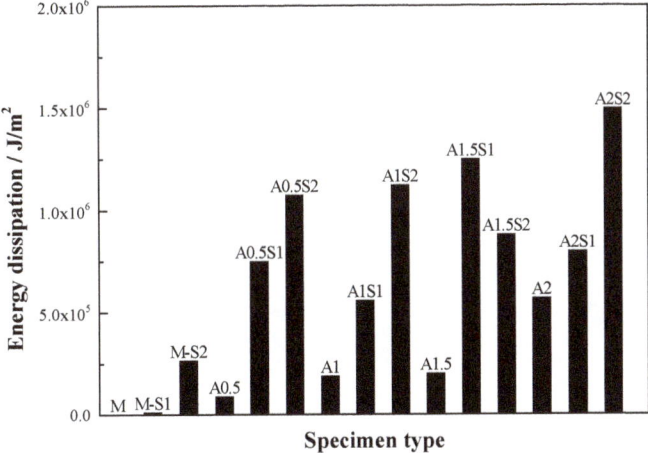

Figure 4. Comparison of energy dissipation of different types of steel grid-PVA fiber reinforced ECC specimens.

Figure 5. Post-failure specimens of steel grid-PVA fiber reinforced ECC with multiple cracks.

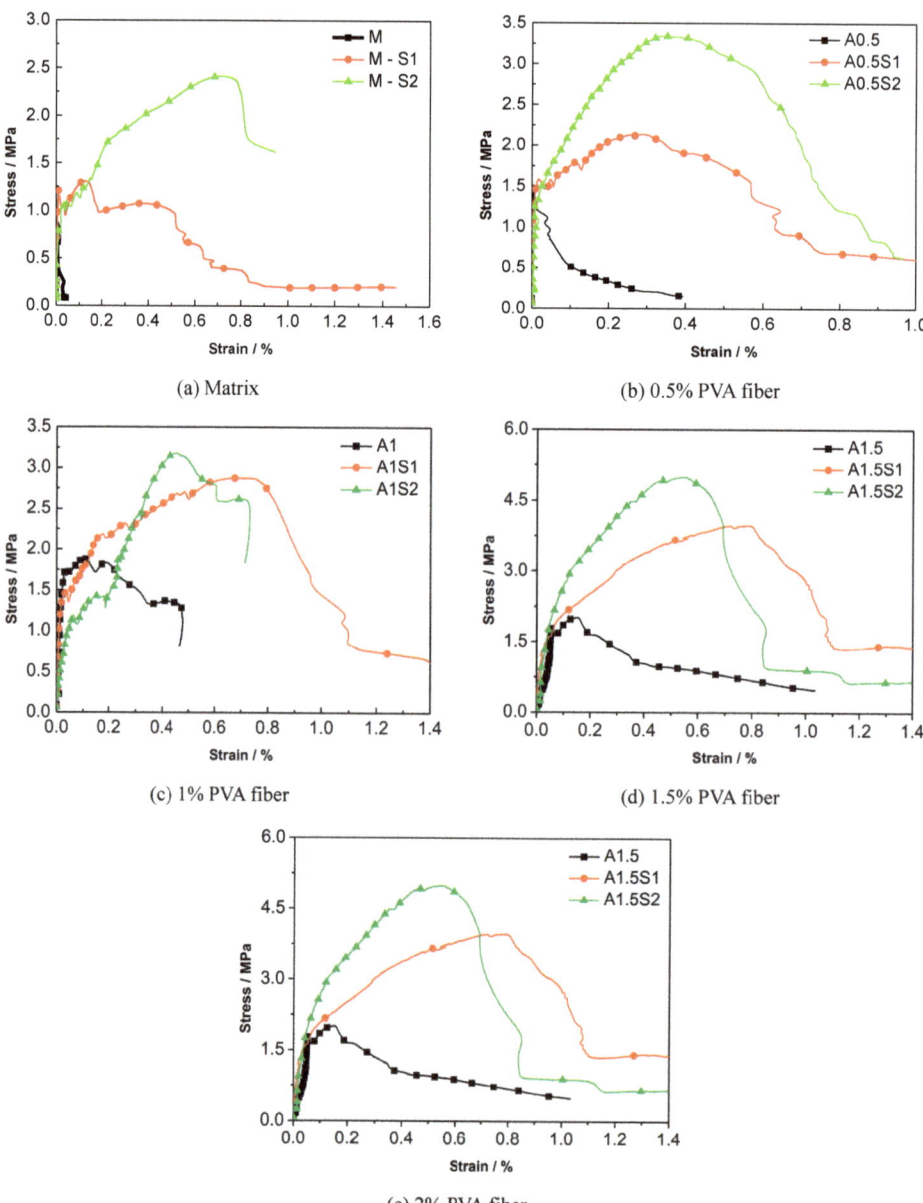

Figure 6. Comparison of stress–strain curves of ECC specimens with different numbers of steel-grid layers. (**a**) Matrix; (**b**) 0.5% PVA fiber; (**c**) 1% PVA fiber; (**d**) 1.5% PVA fiber; (**e**) 2% PVA fiber.

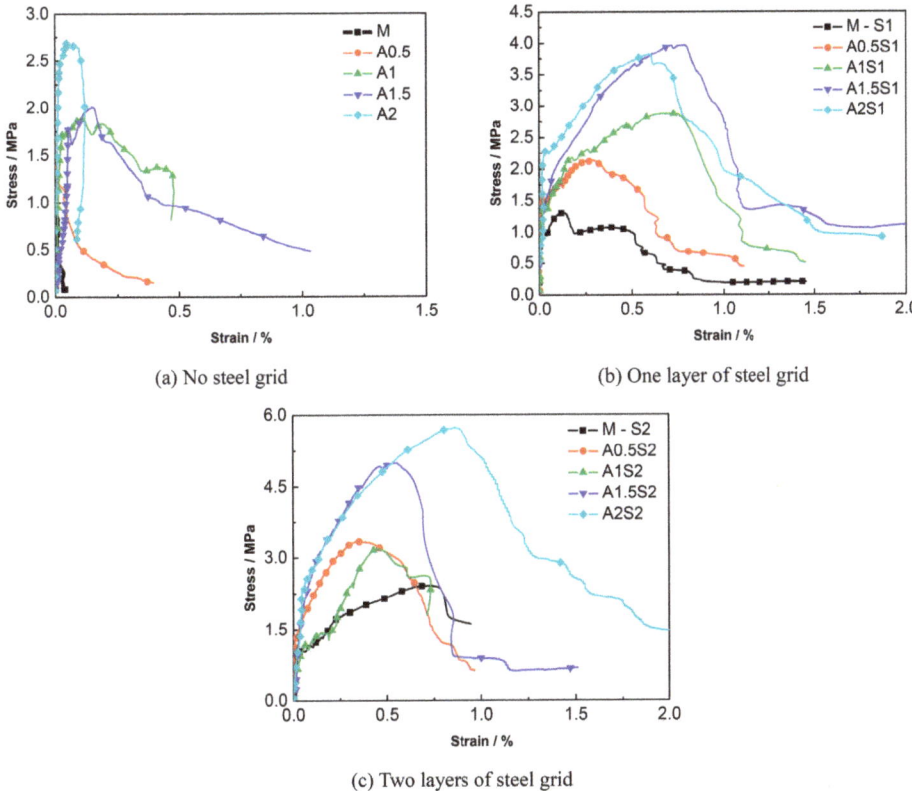

Figure 7. Comparison of stress-strain curves of ECC specimens with different PVA fiber volume fractions. (**a**) No steel grid; (**b**) one layer of steel grid; (**c**) two layers of steel grid.

3.2. Quasi-Static Tensile Test Results of the Steel Grid-KEVLAR Fiber Reinforced ECC

The quasi-static tensile test results of the ECC reinforced with steel grids and KEVLAR fibers are presented in Table 7. In the specimen notation, "M" stands for the matrix, "S" stands for the steel grid, "K" stands for the KEVLAR fibers, and the two numbers stand for the volume content of the KEVLAR fibers and the number of steel-grid layers, respectively. For example, "K1S2" stands for the ECC specimen with a volume content of the KEVLAR fibers of 1% and two layers of steel grid.

The comparison of the energy dissipation of different types of steel grid-KEVLAR fiber reinforced specimens is shown in Figure 8. It is illustrated that the energy dissipation performance of the ECC can be improved remarkably by the addition of a steel grid. When two layers of steel grid are added to the matrix, the energy dissipation performance increases by about two to eight times.

A comparison of the stress-strain curves of the specimens with different numbers of steel-grid layers is shown in Figure 9. A comparison for the specimens with different fiber volume fractions is shown in Figure 10. The results indicate that the tensile strength of the KEVLAR-ECC can be enhanced by the addition of steel grids. For the fiber volume fraction of 1.0%, the maximal peak tensile stress increase of about 50% or 140% compared to the matrix specimen by adding one layer or two layers of steel grid can be obtained, respectively. In addition, the ductility of the KEVLAR-ECC can also be improved noticeably by the addition of steel grids. For the KEVLAR-ECC specimens, the ultimate strain increases by about two or 3.5 times compared to the matrix specimen by adding one layer or two layers of steel grid, respectively. The peak tensile stress of the steel grid-KEVLAR fiber reinforced ECC can be enhanced with the increase of the volume fraction of KEVLAR fibers. For the volume fractions of 0.5%, 1%, 1.5%, and 2%, the peak tensile stress increases by about 80%, 65%, 125%, and 200% compared to the matrix specimen, respectively. The tensile strength of the KEVLAR-ECC has a maximum average increase when the volume fraction of the KEVLAR fibers is 2%. But relative to the PVA-ECC, the ductility of the KEVLAR-ECC is poor for the volume fractions of 1.5% and 2%. For instance, the ultimate strain of K1.5S2 is 70% of that of A1.5S2, the ultimate strain of K2S1 is about 52% of that of A2S1. This may be because the surface of KEVLAR fiber has not been adequately cleaned, and this leads to a relatively weak bond between the KEVLAR fiber and matrix.

Table 7. Quasi-static tensile test results of the steel grid-KEVLAR fiber reinforced ECC.

Specimen Type	Initial Cracking Stress (MPa)	Peak Stress (MPa)	Ultimate Strain (%)	Energy Dissipation (J/m^2)
M	0.98	0.98	0.01	1578
M-S1	0.86	1.29	0.72	11,896
M-S2	1.77	2.33	0.32	265,793
K0.5	1.79	1.79	0.06	49,591
K0.5S1	1.93	2.07	0.16	464,231
K0.5S2	3.28	3.28	0.55	1,099,560
K1	1.6	1.6	0.12	168,336
K1S1	1.88	2.45	0.42	345,744
K1S2	3.81	3.81	0.59	1,314,334
K1.5	2.21	2.21	0.26	343,120
K1.5S1	2.25	2.76	0.22	357,616
K1.5S2	4.48	4.88	0.45	639,744
K2	2.93	2.93	0.24	332,010
K2S1	3.05	3.39	0.35	280,896
K2S2	4.15	4.15	0.29	884,450

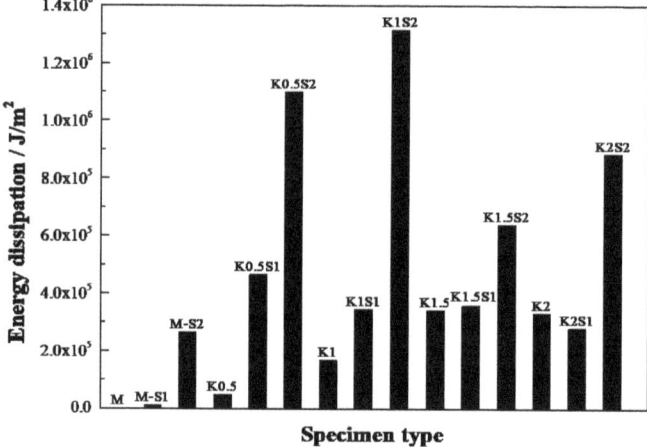

Figure 8. Comparison of energy dissipation of different types of steel grid-KEVLAR fiber reinforced ECC specimens.

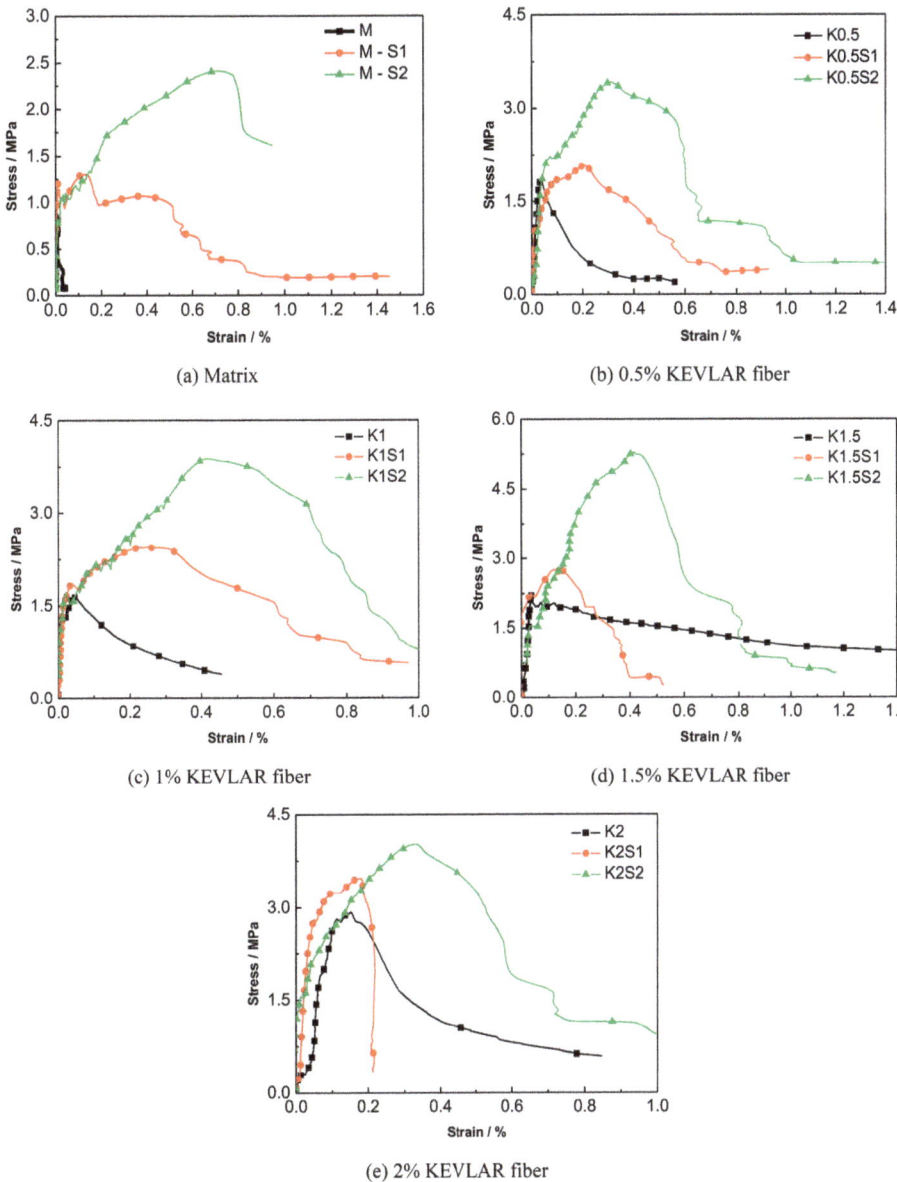

Figure 9. Comparison of stress-strain curves of ECC specimens with different numbers of steel-grid layers. (**a**) Matrix; (**b**) 0.5% KEVLAR fiber; (**c**) 1% KEVLAR fiber; (**d**) 1.5% KEVLAR fiber; (**e**) 2% KEVLAR fiber.

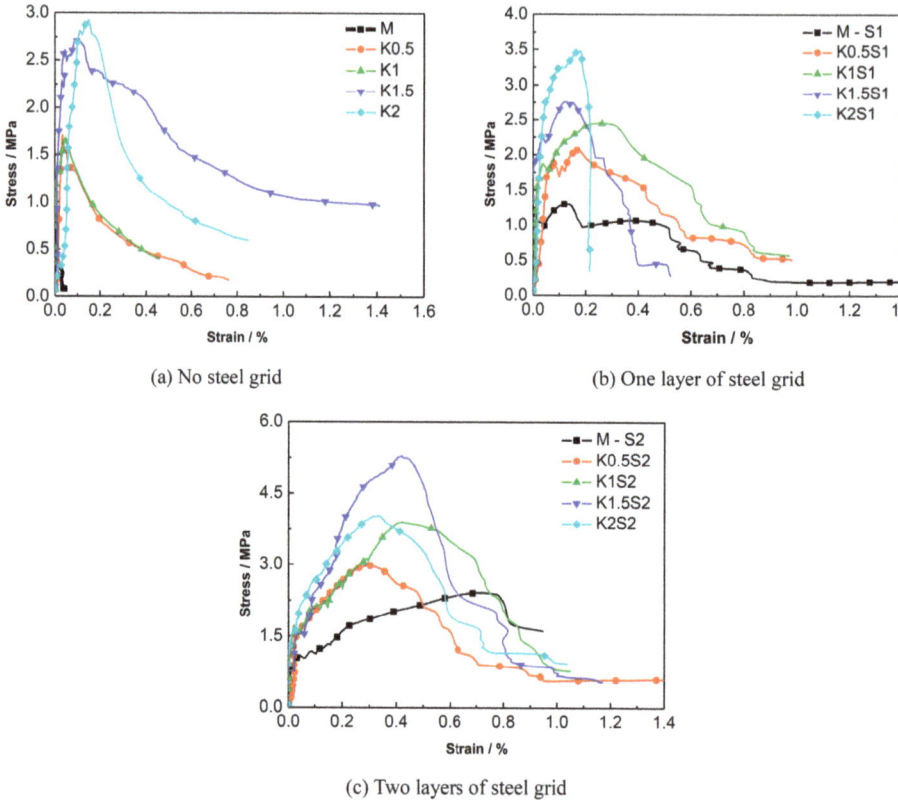

Figure 10. Comparison of stress-strain curves of ECC specimens with different KEVLAR fiber volume fractions. (**a**) No steel grid; (**b**) One layer of steel grid; (**c**) Two layers of steel grid.

3.3. Quasi-Static Tensile Test Results of the Steel Grid-PEFiber Reinforced ECC

The quasi-static tensile test results of the ECC reinforced with steel grids and PE fibers are presented in Table 8. In the specimen notation, "M" stands for the matrix, "S" stands for the steel grid, "E" stands for the PE fiber, and the two numbers stand for the volume content of the PE fibers and the number of steel-grid layers, respectively. For example, "E2S2" stands for the ECC specimen with a volume content of the PE fibers of 2% and two layers of steel grid.

A comparison of the energy dissipation of different types of steel grid-PE fiber reinforced specimens is shown in Figure 11. It is shown that the PE-ECC with a fiber volume fraction of 1–2% exhibits excellent energy dissipation performance. A comparison of the stress-strain curves of the specimens with different numbers of steel-grid layers is shown in Figure 12. A comparison for the specimens with different fiber volume fractions is shown in Figure 13. The results indicate that the tensile strength of the PE-ECC can be enhanced by the addition of steel grids. For the fiber volume fraction of 0.5%, the maximal peak tensile stress increase of about 80% or 190% compared to the matrix specimen by adding one layer or two layers of steel grid can be obtained, respectively. On the other hand, the ductility of the PE-ECC has not been improved remarkably by the addition of steel grids. For the specimens with a fiber volume fraction of 1–2%, the ultimate strain decreases when the steel grid is added to the matrix. At the initial stage of tensile loading, the steel grid has a strong bond with the matrix. After the peak stress, the steel grid will separate from the matrix, and this leads to the deformation inconsistency between the steel grid and the matrix. The results indicate that the ultimate strain mainly depends on the deformation capacity of the steel grid. The deformation capacity of the steel grid is, however, lower than that of the PE-ECC with a relatively high fiber volume fraction, such as 1.5% or 2%. The PE-ECC without steel-grid reinforcement shows a ductile failure behavior. The ductility and energy dissipation performance can be improved with the increase of the volume fraction of PE fibers. The peak tensile stress of the steel grid-PE fiber reinforced ECC can be enhanced with the increase of the volume fraction of PE fibers. For the volume fraction of 0.5%, 1%, 1.5%, and 2%, the peak tensile stress increases by about 70%, 80%, 130%, and 160% compared to the matrix specimen, respectively.

Table 8. Quasi-static tensile test results of the steel grid-PE fiber reinforced ECC.

Specimen Type	Initial Cracking Stress (MPa)	Peak Stress (MPa)	Ultimate Strain (%)	Energy Dissipation (J/m^2)
M	0.98	0.98	0.01	1578
M-S1	0.86	1.29	0.72	11,896
M-S2	1.77	2.33	0.32	265,793
E0.5	1.21	1.21	0.24	347,384
E0.5S1	1.35	2.19	1.15	381,290
E0.5S2	1.52	3.59	0.50	556,080
E1	1.01	1.40	1.98	1,300,963
E1S1	1.17	2.32	0.4	8278.26
E1S2	3.60	3.60	0.51	436,100
E1.5	1.64	1.94	4.87	3,284,074
E1.5S1	2.67	2.97	0.58	142,450
E1.5S2	3.96	3.96	0.58	422,240
E2	1.8	2.18	5.18	3,284,074
E2S1	3.24	3.33	1.02	335,440
E2S2	4.34	4.34	0.79	432,075

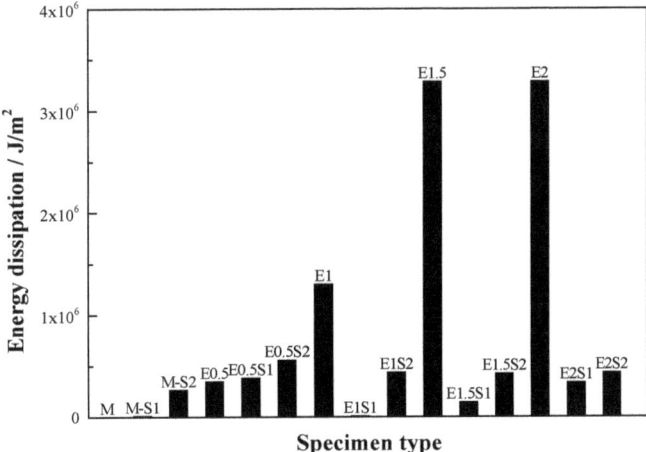

Figure 11. Comparison of energy dissipation of different types of steel grid-PE fiber reinforced ECC specimens.

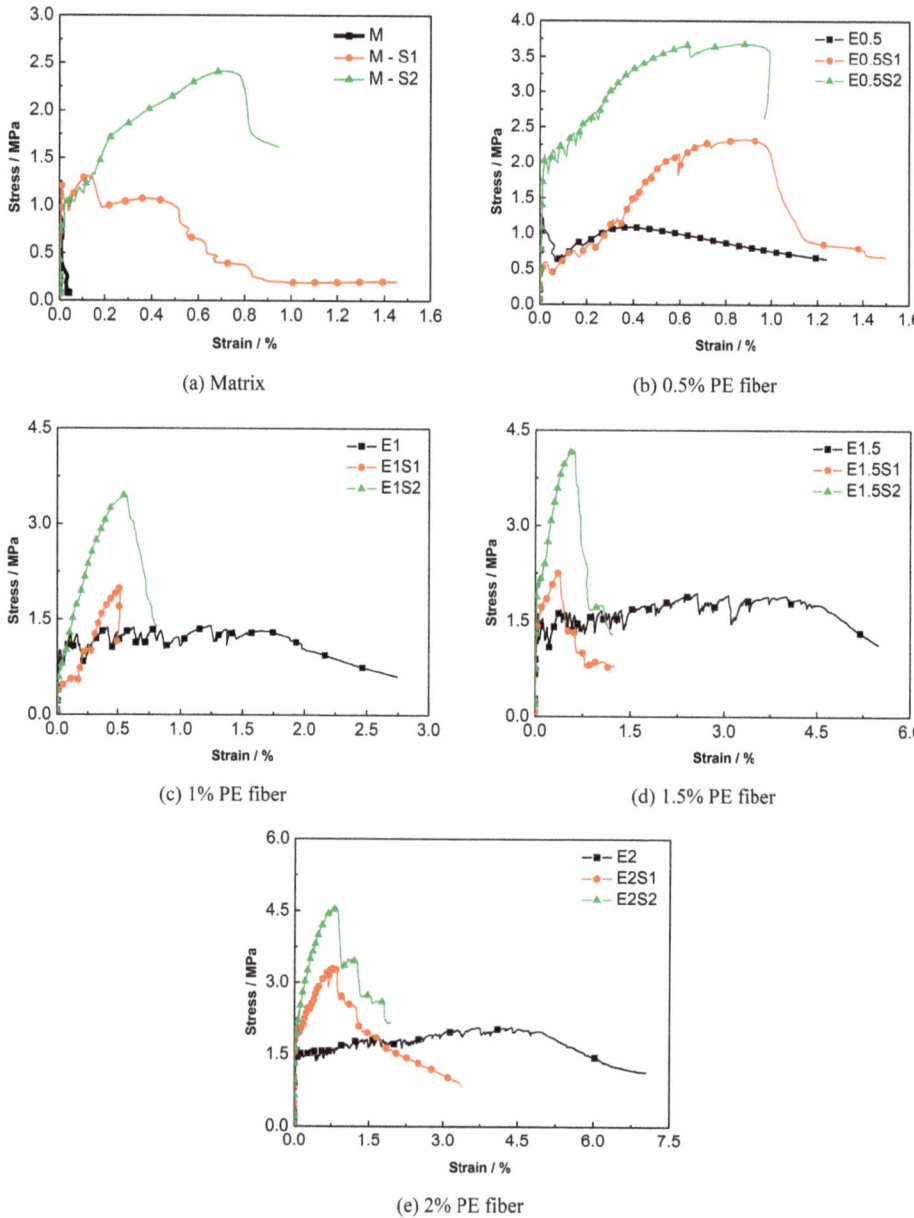

Figure 12. Comparison of stress-strain curves of ECC specimens with different numbers of steel-grid layers. (**a**) Matrix; (**b**) 0.5% PE fiber; (**c**) 1% PE fiber; (**d**) 1.5% PE fiber; (**e**) 2% PE fiber.

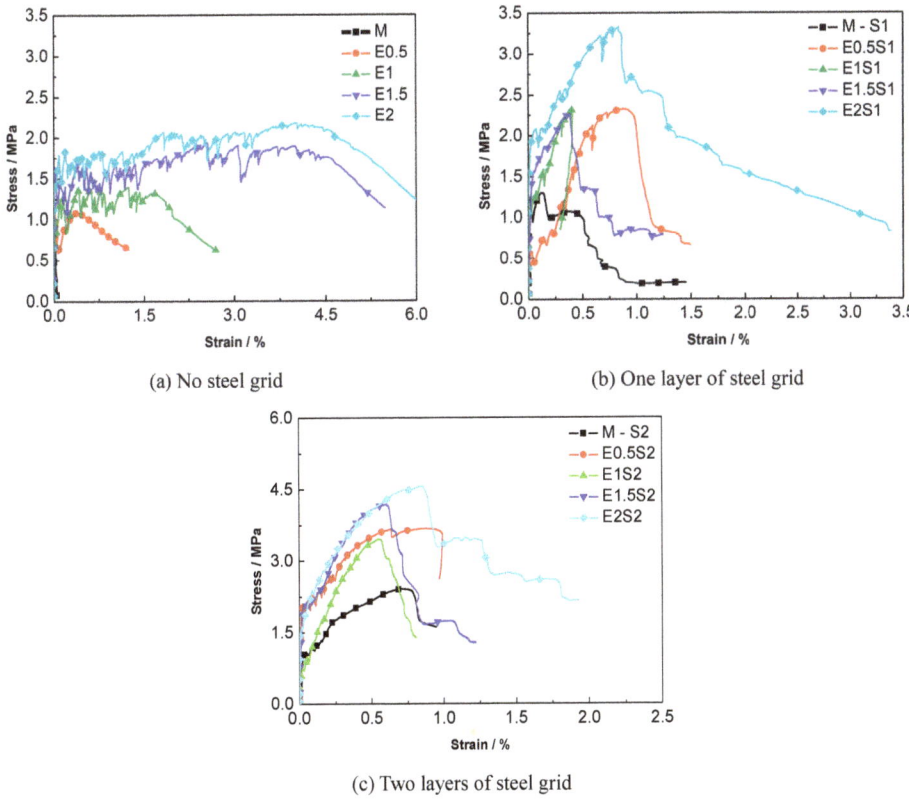

Figure 13. Comparison of stress-strain curves of ECC specimens with different PE fiber volume fractions. (**a**) No steel grid; (**b**) One layer of steel grid; (**c**) Two layers of steel grid.

4. Mechanical Model for Quasi-Static Tensile Strength of Steel Grid-Fiber Reinforced ECC

In this section, a mechanical model for the quasi-static initial and ultimate tensile strength of the steel grid-fiber reinforced ECC is proposed. The initial tensile strength corresponds to the tensile stress when the first crack appears in the specimen. The steel grid-fiber reinforced ECC is treated as a composite material consisting of steel grid and ECC matrix. The total tensile strength of the steel grid-fiber reinforced ECC can be acquired from the sum of the tensile strength of steel grid and that of ECC matrix.

4.1. Model Development

The stress-strain relation of the steel grid for the tensile loading can be treated as an ideal hyperbolic model as shown in Figure 14. It can be expressed as Equations (3) and (4):

$$\sigma_s = \varepsilon E_s \quad \varepsilon < \varepsilon_y. \tag{3}$$

$$\sigma_s = \sigma_y + (\varepsilon - \varepsilon_y)E_t \quad \varepsilon_y \leq \varepsilon \leq \varepsilon_u. \tag{4}$$

where σ_s and ε are the tensile stress and strain of the steel grid, respectively. σ_y is the yield stress of the steel grid, ε_y and ε_u are the yield strain and ultimate strain of the steel grid, respectively. E_s and E_t are the elastic modulus of the steel grid at the pre-yield and post-yield stage, respectively.

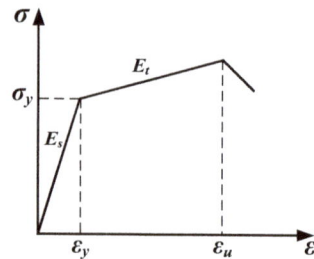

Figure 14. Stress-strain relation of steel grid for the tensile loading.

The tensile strength of ECC matrix consists of the strength related to the crack end toughness and that related to the fiber bonding. From the fracture mechanics theory, the stress strength factor of a straight crack under the action of tensile loads can be expressed as Equation (5) [22]:

$$K_L = \sigma_L \sqrt{2w \tan\left(\frac{\pi c}{2w}\right)} \quad (5)$$

where w is the width of the specimen, c is the length of the crack, and σ_L is the tensile stress acting on the ECC matrix. For a composite material, the stress strength factor K_L equals the crack end toughness of the composite material K_{tip}. The tensile stress σ_L can, thus, be written as Equation (6):

$$\sigma_L = \frac{K_{tip}}{\sqrt{2w \tan\left(\frac{\pi c}{2w}\right)}} \quad (6)$$

The crack end toughness of the composite material K_{tip} can be expressed as Equation (7) [23]:

$$K_{tip} = \frac{E_c}{E_m} K_m \quad (7)$$

where E_c and E_m are the elastic modulus of the composite material and matrix, respectively. K_m is the crack end toughness of the matrix.

For a single fiber in the ECC matrix, the relation between bonding load P_f and crack split displacement δ during the debonding process can be expressed as Equations (8) and (9) [24]:

$$P_f(\delta) = \frac{\pi}{2}\sqrt{(1+\eta)E_f d_f^3 \tau \delta}e^\phi \quad \delta \leq \delta_0 \quad (8)$$

$$P_f(\delta) = \pi \tau l d_f \left(1 - \frac{\delta}{l}\right)e^\phi \quad \delta_0 < \delta \leq l \quad (9)$$

where $\delta_0 = 4l^2\tau/(1+\eta)E_f d_f$ is the crack split displacement of the fiber with a length of l when full debonding occurs. d_f is the diameter of the fiber, τ is the bonding strength of the fiber-matrix interface. $\eta = V_f E_f / V_m E_m$, where V_f and V_m are the volume fractions of fiber and matrix, respectively, and E_f is the elastic modulus of the fiber. ϕ is the angle between fiber orientation and the acting direction of the bonding load P_f.

The total fiber bonding stress of the fiber-reinforced ECC matrix can be obtained by the integration of bonding load P_f. It can be written as Equation (10) [25]:

$$\sigma_B(\delta) = \frac{4V_f}{\pi d_f^2} \int_{\phi=0}^{\pi/2} \int_{z=0}^{(l\cos\phi)/2} P_f(\delta) p(\phi) p(z) dz d\phi \quad (10)$$

where $p(\phi)$ and $p(z)$ are the probability density functions of fiber orientation angle ϕ and fiber centroid position z, respectively. For three-dimensional randomly distributed fibers, these two probability density functions are expressed as Equations (11) and (12):

$$p(z) = \frac{2}{l} \quad 0 \leq z \leq \frac{l}{2}\cos\phi \tag{11}$$

$$p(\phi) = \sin\phi \quad 0 \leq \phi \leq \frac{\pi}{2} \tag{12}$$

Substituting Equations (8), (11), and (12) into Equation (10) results in the expression of the total fiber bonding stress of fiber-reinforced ECC before the full debonding as follows Equation (13):

$$\sigma_B(\delta) = \sigma_0 g \left[2\left(\frac{\delta}{\delta_*}\right)^{1/2} - \frac{\delta}{\delta_*} \right] \quad \delta \leq \delta_* \tag{13}$$

where $\delta_* = l^2 \tau / (1+\eta) E_f d_f$ is the crack split displacement of the fiber with a length of $l/2$ when full debonding occurs, $\sigma_0 = \tau V_f l / 2 d_f$, and g is the buffer factor and can be expressed as Equation (14):

$$g = \frac{2}{4+f^2}\left(1 + e^{\pi f/2}\right) \tag{14}$$

where f is the buffer coefficient. In the current study, $f = 0.8$ [26].

The crack split displacement δ can be expressed as Equation (15) [23]:

$$\delta = \delta_a \sqrt{c\left(1 - \frac{r^2}{c^2}\right)} \tag{15}$$

where $\delta_a = 2K_m(1-v^2)/E_m \sqrt{\pi}$, and v is the Poisson ratio of the ECC matrix.

Substituting Equation (15) into Equation (13) results in the bonding stress of a single fiber, and the total fiber bonding force on the crack can be expressed as Equation (16):

$$F_B = 2\int_0^c \sigma_B(\delta) dr \tag{16}$$

F_B corresponds to a specimen width of $2w$. The integration of Equation (16) results in the expressions of the tensile stress related to fiber bonding as follows Equations (17) and (18):

$$\sigma_F = \frac{c}{w}\sigma_0 g\left[1.748 \times \sqrt{\frac{\delta_a}{\delta_*}}(c)^{1/4} - \frac{\pi}{4}\frac{\delta_a}{\delta_*}\sqrt{c}\right] \quad c \leq w \tag{17}$$

$$\sigma_F = \sigma_0 g\left[1.748 \times \sqrt{\frac{\delta_a}{\delta_*}}(c)^{1/4} - \frac{\pi}{4}\frac{\delta_a}{\delta_*}\sqrt{c}\right] \quad c > w \tag{18}$$

Taking both the crack end toughness and fiber bonding into consideration results in the tensile stress-crack split displacement relation as follows Equations (19) and (20):

$$\sigma_c = \frac{K_{tip}}{\sqrt{2w\tan\left(\frac{\pi c}{2w}\right)}} + \frac{c}{w}\sigma_0 g\left[1.748 \times \sqrt{\frac{\delta_a}{\delta_*}}(c)^{1/4} - \frac{\pi}{4}\frac{\delta_a}{\delta_*}\sqrt{c}\right] \quad c < w \tag{19}$$

$$\sigma_c = \sigma_0 g\left[1.748 \times \sqrt{\frac{\delta_a}{\delta_*}}(c)^{1/4} - \frac{\pi}{4}\frac{\delta_a}{\delta_*}\sqrt{c}\right] \quad c \geq w \tag{20}$$

When the length of the crack c is greater than the width of the specimen w, the crack end toughness is not taken into consideration, and the K_{tip} term is deleted from the tensile stress-crack split displacement relation. Letting $c = w$, we can obtain the initial tensile strength of the ECC matrix as follows Equation (21):

$$\sigma_{fc} = \frac{K_{tip}}{\sqrt{2w}} + \sigma_0 g \left[1.748 \times \sqrt{\frac{\delta_a}{\delta_*}} (w)^{1/4} - \frac{\pi}{4} \frac{\delta_a}{\delta_*} \sqrt{w} \right] \quad (21)$$

For the tensile ultimate state of the ECC matrix, the crack end toughness is neglected. Letting $\delta = \delta^*$, we can obtain the ultimate tensile strength of the ECC matrix as follows Equation (22):

$$\sigma_{tu} = 0.963 \times \sigma_0 g \quad (22)$$

The initial tensile strength of steel grid-fiber reinforced ECC can be obtained by the sum of the initial tensile strength of the steel grid and that of the ECC matrix. It can be expressed as follows Equation (23):

$$\sigma_{fc} = \frac{F_{Sc}}{A_c} + \frac{K_{tip}}{\sqrt{2w}} + \sigma_0 g \left[1.748 \times \sqrt{\frac{\delta_a}{\delta_*}} (c)^{1/4} - \frac{\pi}{4} \frac{\delta_a}{\delta_*} \sqrt{c} \right] \quad (23)$$

where F_{Sc} is the tensile force carried by steel grid when the first crack appears in the specimen, and A_c is the cross-sectional area of specimen.

The ultimate tensile strength of steel grid-fiber reinforced ECC can be expressed as follows Equation (24):

$$\sigma_{tu} = \frac{F_{Su}}{A_c} + 0.963 \times \sigma_0 g \quad (24)$$

where F_{Su} is the ultimate tensile force of the steel grid.

4.2. Model Validation

The mechanical model for the quasi-static initial and ultimate tensile strength of the steel grid-fiber reinforced ECC proposed in the current study is validated by the quasi-static tension test data of the steel grid-PE fiber reinforced ECC. The comparisons of the analytical and experimental initial and ultimate tensile strengths of the steel grid-PE fiber reinforced ECC are presented in Tables 9 and 10, respectively.

It is shown that, as a whole, the analytical initial tensile strength results predicted by the proposed mechanical model agree well with the corresponding experimental results, with most error below 20%. The analytical ultimate tensile strength results predicted by the proposed mechanical model are, however, larger than the corresponding experimental results. This means that the proposed mechanical model overestimates the ultimate tensile strength of the steel grid-fiber reinforced ECC. This is because the slip and debonding between the steel grid and the ECC matrix during the tensile loading are not taken into consideration in the proposed mechanical model. The tensile stress of steel grid increases after the peak stress in the proposed model, which is contrary to the observed test behavior of the steel grid. In the tensile tests, the tensile stress of the steel grid decreases after the peak stress.

Table 9. Comparison of analytical and experimental initial tensile strengths of steel grid-PE fiber reinforced ECC.

Specimen Type	Analytical Initial Tensile Strength (MPa)	Experimental Initial Tensile Strength (MPa)	Error (%)
E0.5S1	1.18	1.35	−12.6
E0.5S2	1.60	1.52	5.3
E1S1	2.09	1.17	78.6
E1S2	2.77	3.60	−23.1
E1.5S1	2.77	2.67	3.7
E1.5S2	3.44	3.96	−13.1
E2S1	3.43	3.24	5.9
E2S2	4.11	4.34	−5.3

Table 10. Comparison of analytical and experimental ultimate tensile strengths of steel grid-PE fiber reinforced ECC.

Specimen Type	Analytical Ultimate Tensile Strength (MPa)	Experimental Ultimate Tensile Strength (MPa)	Error (%)
E0.5S1	2.86	2.19	30.6
E0.5S2	4.73	3.59	31.8
E1S1	3.84	2.32	65.5
E1S2	5.71	3.60	58.6
E1.5S1	4.83	2.97	62.6
E1.5S2	6.70	3.96	69.2
E2S1	5.81	3.33	74.5
E2S2	7.68	4.34	77.0

Note: Error = (analytical results−experimental results)/experimental results × 100%.

5. Conclusions

In this research, an engineered cementitious composite (ECC) was reinforced with steel grids and different types of fibers to improve its tensile strength and ductility. A series of tensile tests have been carried out to investigate the quasi-static tensile capacity of the reinforced ECC using a Z100 material tensile testing machine manufactured by the Zwick/Roell Group of Germany. The quasi-static tensile capacity of reinforced ECCs with different numbers of steel-grid layers, types of fibers (PVA fibers, KEVLAR fibers, and PE fibers), and volume fractions of fibers have been tested and compared. It is indicated by the test results that:

(1). On the whole, the steel grid-PVA fiber, steel grid-KEVLAR fiber, and steel grid-PE fiber reinforced ECCs all have higher tensile strength and ductility than the ECC matrix. They can also exhibit excellent energy dissipation performance.

(2). The ultimate tensile strength of the reinforced ECC can be improved by the addition of steel grids. For the steel grid-PVA fiber reinforced ECC, when the fiber volume fraction is 1.5%, a maximal peak tensile stress increase of about 95% or 160% compared to the matrix specimen by adding one layer or two layers of steel grid can be obtained, respectively. For the steel grid-KEVLAR fiber reinforced ECC, when the fiber volume fraction is 1.0%, the maximal peak tensile stress increase of about 50% or 140% compared to the matrix specimen by adding one layer or two layers of steel grid can be obtained, respectively. For the steel grid-PE fiber reinforced ECC, when the fiber volume fraction is 0.5%, the maximal peak tensile stress increase of about 80% or 190% compared to the matrix specimen by adding one layer or two layers of steel grid can be obtained, respectively. A more remarkable increase of ultimate tensile strength can be obtained by adding two layers of steel grid, but it is more difficult to have a firm bonding with the ECC matrix for two layers of steel grid than one layer of steel grid.

(3). The ultimate tensile strength of the reinforced ECC can be enhanced with the increase of fiber volume fraction. For all of the fiber types investigated, a volume fraction between 1.5% and 2% can make the reinforced ECC gain the best tensile strength. With these higher fiber volume fractions, the reinforced ECC exhibits strain hardening behavior, and its peak tensile stress increases considerably. The energy dissipation performance of the reinforced ECC can also be enhanced remarkably.

(4). The ductility of PVA fiber reinforced ECC can be improved by the addition of steel grids and the increase of fiber volume fraction. The phenomenon of multiple-cracking can be observed for steel grid-PVA fiber reinforced ECC. The steel grid-PE fiber reinforced ECC also exhibits significant ductility and energy dissipation performance. Under those circumstances when excellent ductility and energy dissipation performance are required, it is better to use PVA fibers or PE fibers. The ductility of the steel grid-KEVLAR fiber reinforced ECC can be improved by the addition of steel grids. The ductility and energy dissipation performance of the steel grid-PE fiber reinforced ECC can be improved with the increase of fiber volume fraction.

A mechanical model for the quasi-static initial and ultimate tensile strengths of the steel grid-fiber reinforced ECC is proposed. This model is validated by the quasi-static tension test data of the steel grid-PE fiber reinforced ECC. It is indicated by the comparison of the analytical tensile strengths with the corresponding test results that the initial tensile strength predicted by the proposed mechanical model is relatively accurate, but the analytical ultimate tensile strength results predicted by the proposed mechanical model are larger than the corresponding experimental results.

Author Contributions: Methodology, J.W. and L.L.; validation, J.W. and M.W.; investigation, W.W.; data curation, W.L.; Writing—Original Draft preparation, L.L.; Writing—Review and Editing, W.L.

Funding: This research was funded by the Key Fundamental Study Development Project of the People's Republic of China (grant number 2015CB058003) and the Beijing Municipal Natural Science Foundation (grant number 8172010).

Acknowledgments: This research was funded by the Key Fundamental Study Development Project of the People's Republic of China (grant number 2015CB058003) and the Beijing Municipal Natural Science Foundation (grant number 8172010). The financial support from both foundations is gratefully acknowledged.

Conflicts of Interest: The authors declare no conflict of interest.

References

1. Li, V.C. On Engineered Cementitious Composites (ECC). *J. Adv. Concr. Technol.* **2003**, *1*, 215–230. [CrossRef]
2. Tran, T.K.; Kim, D.J. Investigating Direct Tensile Behavior of High Performance Fiber Reinforced Cementitious Composites at High Strain Rates. *Cem. Concr. Res.* **2013**, *50*, 62–73. [CrossRef]
3. Arboleda, D.; Carozzi, F.G.; Nanni, A.; Poggi, C. Testing Procedures for the Uniaxial Tensile Characterization of Fabric-Reinforced Cementitious Matrix Composites. *J. Compos. Constr.* **2016**, *20*, 44–54. [CrossRef]
4. Kim, J.S.; Cho, C.G.; Moon, H.J.; Kim, H.; Lee, S.J.; Kim, W.J. Experiments on Tensile and Shear Characteristics of Amorphous Micro Steel (AMS) Fibre-Reinforced Cementitious Composites. *Int. J. Concr. Struct. Mater.* **2017**, *11*, 647–655. [CrossRef]
5. Ali, M.A.E.M.; Soliman, A.M.; Nehdi, M.L. Hybrid-Fiber Reinforced Engineered Cementitious Composite under Tensile and Impact Loading. *Mater. Des.* **2017**, *117*, 139–149. [CrossRef]
6. Nehdi, M.L.; Ali, M.A.E.M. Experimental and Numerical Study of Engineered Cementitious Composite with Strain Recovery under Impact Loading. *Appl. Sci.* **2019**, *9*, 994. [CrossRef]
7. Yu, K.Q.; Wang, Y.C.; Yu, J.T.; Xu, S.L. A Strain-Hardening Cementitious Composites with the Tensile Capacity up to 8%. *Constr. Build. Mater.* **2017**, *137*, 410–419. [CrossRef]
8. Yu, K.Q.; Dai, J.G.; Lu, Z.D.; Poon, C.S. Rate-Dependent Tensile Properties of Ultra-High Performance Engineered Cementitious Composites (UHP-ECC). *Cem. Concr. Compos.* **2018**, *93*, 218–234. [CrossRef]
9. Curosu, I.; Liebscher, M.; Mechtcherine, V.; Bellmannb, C.; Michelb, S. Tensile Behavior of High-Strength Strain-Hardening Cement-Based Composites (HS-SHCC) Made with High-Performance Polyethylene, Aramid and PBO Fibers. *Cem. Concr. Res.* **2017**, *98*, 71–81. [CrossRef]
10. Zhou, Y.W.; Xi, B.; Yu, K.Q.; Sui, L.L.; Xing, F. Mechanical Properties of Hybrid Ultra-High Performance Engineered Cementitious Composites Incorporating Steel and Polyethylene Fibers. *Materials* **2018**, *11*, 1448. [CrossRef] [PubMed]
11. Zhang, W.; Yin, C.L.; Ma, F.Q.; Huang, Z.Y. Mechanical Properties and Carbonation Durability of Engineered Cementitious Composites Reinforced by Polypropylene and Hydrophilic Polyvinyl Alcohol Fibers. *Materials* **2018**, *11*, 1147. [CrossRef] [PubMed]
12. Kim, M.J.; Kim, S.; Yoo, D.Y. Hybrid Effect of Twisted Steel and Polyethylene Fibers on the Tensile Performance of Ultra-High-Performance Cementitious Composites. *Polymers* **2018**, *10*, 879. [CrossRef] [PubMed]
13. Zhu, Z.F.; Wang, W.W.; Harries, K.A.; Zheng, Y.Z. Uniaxial Tensile Stress-Strain Behavior of Carbon-Fiber Grid-Reinforced Engineered Cementitious Composites. *J. Compos. Constr.* **2018**, *22*, 163–176. [CrossRef]
14. Al-Gemeel, A.N.; Zhuge, Y.; Youssf, O. Experimental Investigation of Basalt Textile Reinforced Engineered Cementitious Composite under Apparent Hoop Tensile Loading. *J. Build. Eng.* **2019**, *23*, 270–279. [CrossRef]

15. Li, B.B.; Xiong, H.B.; Jiang, J.F.; Dou, X.X. Tensile Behavior of Basalt Textile Grid Reinforced Engineering Cementitious Composite. *Compos. Pt. B Eng.* **2019**, *156*, 185–200. [CrossRef]
16. Sun, M.; Chen, Y.Z.; Zhu, J.Q.; Sun, T.; Shui, Z.H.; Ling, G.; Zhong, H.X.; Zheng, Y.R. Effect of Modified Polyvinyl Alcohol Fibers on the Mechanical Behavior of Engineered Cementitious Composites. *Materials* **2019**, *12*, 37. [CrossRef] [PubMed]
17. Wang, Z.B.; Zhang, J.; Wang, J.H.; Shi, J.H. Tensile Performance of Polyvinyl Alcohol-Steel Hybrid Fiber Reinforced Cementitious Composite with Impact of Water to Binder Ratio. *J. Compos. Mater.* **2015**, *49*, 2169–2186.
18. Abrishambaf, A.; Pimentel, M.; Nunes, S. Influence of Fibre Orientation on the Tensile Behaviour of Ultra-High Performance Fibre Reinforced Cementitious Composites. *Cem. Concr. Res.* **2017**, *97*, 28–40. [CrossRef]
19. Wu, H.L.; Yu, J.; Zhang, D.; Zheng, J.X.; Li, V.C. Effect of Morphological Parameters of Natural Sand on Mechanical Properties of Engineered Cementitious Composites. *Cem. Concr. Compos.* **2019**, *100*, 108–119. [CrossRef]
20. Pourfalah, S. Behaviour of Engineered Cementitious Composites and Hybrid Engineered Cementitious Composites at High Temperatures. *Constr. Build. Mater.* **2018**, *158*, 921–937. [CrossRef]
21. Du, Q.; Wei, J.; Lv, J. Effects of High Temperature on Mechanical Properties of Polyvinyl Alcohol Engineered Cementitious Composites (PVA-ECC). *Int. J. Civ. Eng.* **2018**, *16*, 965–972. [CrossRef]
22. Lawn, B. *Fracture of Brittle Solids*, 2nd ed.; Cambridge University Press: New York, NY, USA, 1993; ISBN 0521409721.
23. Marshall, D.B.; Cox, B.N.; Evans, A.G. Mechanics of Matrix Cracking in Brittle-Matrix Fiber Composites. *Acta Metall.* **1985**, *33*, 2013–2021. [CrossRef]
24. Li, V.C.; Leung, C.K.Y. Steady-State and Multiple Cracking of Short Random Fiber Composites. *J. Eng. Mech.* **1992**, *118*, 2246–2264. [CrossRef]
25. Li, V.C.; Wang, Y.; Backer, S. A Micromechanical Model of Tension Softening and Bridging Toughening of Short Random Fiber Reinforced Brittle Matrix Composites. *J. Mech. Phys. Solids* **1991**, *39*, 607–625. [CrossRef]
26. Kanda, T.; Li, V.C. A New Micromechanics Design Theory for Pseudo Strain Hardening Cementitious Composite. *J. Eng. Mech. ASCE* **1999**, *125*, 373–381. [CrossRef]

© 2019 by the authors. Licensee MDPI, Basel, Switzerland. This article is an open access article distributed under the terms and conditions of the Creative Commons Attribution (CC BY) license (http://creativecommons.org/licenses/by/4.0/).

Article

Evolution of Rheological Behaviors of Styrene-Butadiene-Styrene/Crumb Rubber Composite Modified Bitumen after Different Long-Term Aging Processes

Yangsheng Ye [1], Gang Xu [2], Liangwei Lou [1], Xianhua Chen [2,*], Degou Cai [1] and Yuefeng Shi [1]

1. Railway Engineering Research Institute, China Academy of Railway Sciences Corporation Limited, Beijing 100081, China
2. School of Transportation, Southeast University, Nanjing 211189, China
* Correspondence: chenxh@seu.edu.cn

Received: 30 May 2019; Accepted: 20 July 2019; Published: 24 July 2019

Abstract: In this study, a new type of composite modified bitumen was developed by blending styrene-butadiene-styrene (SBS) and crumb rubber (CR) with a chemical method to satisfy the durability requirements of waterproofing material in the waterproofing layer of high-speed railway subgrade. A pressure-aging-vessel test for 20, 40 and 80 h were conducted to obtain bitumen samples in different long-term aging conditions. Multiple stress creep recovery (MSCR) tests, linear amplitude scanning tests and bending beam rheometer tests were conducted on three kinds of asphalt binders (SBS modified asphalt, CR modified asphalt and SBS/CR composite modified asphalt) after different long-term aging processes, including high temperature permanent deformation performance, resistance to low temperature thermal and fatigue crack. Meanwhile, aging sensitivities were compared by different rheological indices. Results showed that SBS/CR composite modified asphalt possessed the best properties before and after aging. The elastic property of CR in SBS/CR composite modified asphalt improved the ability to resist low temperature thermal and fatigue cracks at a range of low and middle temperatures. Simultaneously, the copolymer network of SBS and CR significantly improved the elastic response of the asphalt SBS/CR modified asphalt at a range of high temperatures. Furthermore, all test results indicated that the SBS/CR modified asphalt possesses the outstanding ability to anti-aging. SBS/CR is an ideal kind of asphalt to satisfy the demand of 60 years of service life in the subgrade of high speed railway.

Keywords: high speed railway; SBS/CR modified asphalt; long-term aging; anti-aging

---

1. Introduction

Waterproofing layers are essential to preventing surface water from infiltrating into a high-speed railway subgrade, which can ensure its stability and bearing capacity, and especially prevent subgrade frost in seasonally frozen regions [1–4]. The dense-graded asphalt concrete was used as a waterproofing material to substitute for fiber-reinforced concrete, and its requirements were proposed based on the practice achievements in the Beijing–Zhangjiakou high-speed railway test section. Meanwhile, theoretical analysis, finite element calculation and past engineering experience show low temperature thermal and fatigue cracking, permanent deformation and passive stretching near the expansion joint of the base are the main failure modes of the asphalt concrete waterproof sealing layer [5–7], therefore, the ability for low temperature thermal and fatigue crack reduction and better deformation recovery are a prerequisite to extending the service life. It must be noted that according to high-speed railway standards in China, the design life of waterproofing layers in the high-speed railway is 60 years, which is four times the service life of the freeway. As is known, aging has a dramatic impact on the aforementioned performance of the asphalt binder or asphalt mixture [8–11].

However, conventional base asphalt is an easily aged material. Meanwhile, the dense-graded asphalt concrete used as waterproofing layers requires distinguished resistance to low-temperature crack, fatigue crack and permanent deformation. One of the most effective methods to achieve a better engineering performance of asphalt binders and mixtures to extend service life is to modify asphalts by specialized refining practices, chemical reactions and additives [12,13]. For instance, styrene-butadiene-styrene (SBS) modifier can improve permanent deformation at the range of the high-temperature domain and anti-aging performance of asphalt binders [14]. Crumb Rubber modified bitumen (CRMB) at high additive content decreases non recoverable deformation and enhances the high elastic response [15,16]. Montmorillonite or wood lignin modified asphalt binders are able to potentially delay the aging process and enhance deformation and fatigue resistance [17]. In recent years, a large number of researchers have begun to attach attention to compound modified asphalt and tend to combine the advantages of different modifiers [18]. For instance, the addition of crumb rubber (CR) and SBS can make an obvious improvement to asphalt in temperature sensitivity and the viscoelastic response behavior [19]. Ageing indexes of conventional parameters are greatly lessoned in SBS modified bitumen (SBSMB) by using the carbon nano-tubes [11].

The SBS/CR composite modification method is most commonly utilized among them. Because it can not only combine the advantages of CR and SBS to achieve a more satisfactory performance, but it can also cut back on environmental pollution using scrap tires [20]. In a number of research studies that studied the best preparation technology of SBS/CR modified bitumen (SBS/CRMB), it has been found that SBS/CRMB could raise high-temperature stability and possess the best aging resistance as SBS and rubber powder modified bitumen [11,18,19]. The SBS/CRMB with SBS and rubber powder in the end obtained the best preparation process reported by Wang et al. [21]. Inorganic or organic powders such as rubber powder, nano-TiO2 and carbon black can be added in order to raise the anti-aging ability of bitumen. However, the SBS/CRMB with a high performance also faced the great challenges of performances deterioration, which was caused by aging of asphalt materials, especially for the thermal oxygen aging and led to the weak viscoelastic performance of asphalt binder [9,10].

The objectives of this study were to select an ideal bitumen of asphalt concrete waterproofing layers. Therefore, in this research a new type of SBS/CRMB with a chemical modification method was developed. Dynamic Shear Rheological (DSR) tests and a Bending Beam Rheometer (BBR) test were conducted on SBS/CRMB before and after the short-term aging of thin film oven test (TFOT) and long-term aging of a pressurized aging vessel (PAV) aging for 20 h, 40 h and 80 h. Then, the rheological properties and the anti-aging properties of SBS/CRMB were evaluated.

2. Materials and Testing Methodology

2.1. Materials and Preparation

The base binder used in this study was provided by SK Co., Ltd., Korea (SK-70, PG 64-22) and the properties are shown in Table 1. SBS and crumb rubber modifier were provided by Jiangsu Baoli International Investment Co., Ltd. (Wuxi, China). Rubber processing oil, which is rich in aromatic and saturates was utilized to make crumb rubber and SBS swell sufficiently in bitumen. The content of rubber processing oil was 4% by mass of base asphalt. Sulfur powder acted as the cross-linking agent and the amount was 0.2% in bitumen weight. The content of SBS and crumb rubber in the SBS/CR composite modified binder (SBS/CRMB) were 5.5% and 15% in base asphalt weight, respectively. The production process of SBS/CR-MB was as follows: Firstly, heat the base asphalt to 165 °C, and then 4% of rubber processing oil and 15% of CR powder (by the mass of base asphalt), which was devulcanization in the laboratory and was added to the base asphalt and then blended by a special double screw extruder. Subsequently, they were sheared by a shearing machine with a rotation speed of 5000 rpm for about 30 min at 180 °C. Secondly, 5.5% wt of SBS was added to the aforementioned mixture, then sheared at a speed of 5000 rpm at 180 °C and the shearing time was observed during the preparation. Thirdly, 0.2% wt sulfur powder was added slowly to the sample at 1000 rpm and blended for about 15 min. Finally, the prepared samples were developed in an oven for 30 min.

Table 1. Properties of base bitumen.

Properties	Unit	Test Results	Test Method
Penetration (25 °C, 100 g, 5 s)	(0.1 mm)	68.9	ASTM D5
softening point (Ring and ball method)	°C	47.2	ASTM D36
Ductility (15 °C, 5 cm/s)	cm	>100	ASTM D113
Change in mass TFOT	%	−0.2	ASTM D2872
Flash point, Cleveland open cup	°C	289	ASTM D92

As mentioned earlier, there were two different types of bitumen which were considered as references for the comparison of SBS/CR composite modified bitumen, namely SBSMB with 5.5% SBS and CRMB with 15% crumb rubber (by the weight of base asphalt). It should be noted that controlling the impact of the preparation on the binder's properties was consistent, the same conditions were utilized except for the modifier during the preparation.

2.2. Test Methods

2.2.1. Aging Method

The asphalt samples were carried out using TFOT aging at 163 °C for 5 h according to ASTM D1754 to simulate the short-term aging of asphalt during the mixing, transportation and paving progress. Meanwhile, the different long-term aged bitumen samples were obtained by conducting TFOT (5 h, 163 °C), followed by a PAV test for 20, 40 and 80 h with the temperature at 100 °C and 2.1 MPa pressure. The purpose of the longer PAV aging times was simply to create a more highly-aged sample and was not aimed to correlate with any expected services life.

2.2.2. Multiple Stress Creep Recovery (MSCR) Test

The MSCR test was conducted as per AASHTO T350 using a dynamic shear rheometer (DSR) device. The diameter of samples used in the testing was 25 mm, the thickness was 1 mm and the testing temperature was 70 °C. The MSCR test result was based on two replicates. The MSCR test consisted of 20 cycles, a 1 s creep period and a 9 s recovery period at a stress level of 0.1 kPa and this was followed by another 10 cycles of creep and recovery at 3.2 kPa according to AASHTO T350. Under two stress conditions (Jnr-diff) the nonrecoverable creep compliance was different and is presented in Equation (1).

$$J_{nr-diff} = \frac{(J_{nr3.2} - J_{nr0.1})}{J_{nr0.1}} \times 100\% \qquad (1)$$

where $J_{nr0.1}$, $J_{nr3.2}$ is the unrecoverable creep compliance at 0.1 kPa and 3.2 kPa (kPa^{-1}), respectively. $J_{nr-diff}$ is the difference in nonrecoverable creep compliance at the two stress levels (%).

2.2.3. Linear Amplitude Scanning (LAS) Test

The LAS test was carried out to characterize the fatigue properties of all bitumen samples under different aging conditions at 25 °C. According to AASHTO TP 101-12, the testing samples in the LAS test were prepared circular with a diameter of 8 mm and a height of 2 mm. Bitumen samples were tested in two stages based on AASHTO TP 101. In the first stage, the frequency sweep test with a strain level of 0.1% was performed with different frequencies (0.2 to 30 Hz and change as a testing table), which was used to obtain the undamaged material parameter (α). In the second stage, a strain sweep test with a strain change from 0.1 to 30% linearly increased at a constant frequency of 10 Hz. The asphalt binder's damage property was analyzed in viscoelastic continuum damage (VECD) mechanics. Three replicates were tested in this paper in order to guarantee the validity of the testing results.

The damage accumulation in the sample in the LAS test was calculated using Equation (2)

$$D(t) \cong \sum_{i=1}^{N} \left[\pi \gamma_0^2 (C_{i-1} - C_i)\right]^{\frac{\alpha}{1+\alpha}} (t_i - t_{i-1})^{\frac{1}{1+\alpha}} \qquad (2)$$

where

$C_{(t)} = \frac{|G^*|(t)}{|G^*|_{initial}}$ = integrity parameter;
G^* = complex shear modulus, MPa;
γ_0 = applied strain, %;
t = testing time, s;
α = 1/m, where m is the slope of the best-fit straight line with log (storage modulus) in vertical axis and log (applied frequency) on the horizontal axis;

$$C_{(t)} = C_0 - C_1 (D(t))^{C_2} \qquad (3)$$

where

C_0 = 1, the initial value of C;
C_1, C_2 = curve-fit coefficients, then change the form as shown below:

$$\lg(C_0 - C_{(t)}) = \lg(C_1) + C_2 \cdot \lg(D(t)) \qquad (4)$$

The value of $D(t)$ at failure, D_f is defined as the $D(t)$ which corresponds to the reduction in initial $|G^*|$ at the peak shear stress. The calculation is as follows:

$$D_f = \left(\frac{C_0 - C_{atpeakstress}}{C_1}\right)^{\frac{1}{C_2}} \qquad (5)$$

where

$C_{atpeakstress} = C_{(t)}$ at peak stress.

The following parameters (A and B) for the binder fatigue performance model can now be calculated and recorded as follows:

$$A = \frac{f(D_f)^k}{k(\pi C_1 C_2)^\alpha} \qquad (6)$$

where

f = loading frequency (10 Hz),
k = 1 + (1 − C_2)α, and
B = 2α.

The binder fatigue performance parameter N_f can now be calculated as follows:

$$N_f = A(\gamma_{max})^B \qquad (7)$$

where:

γ_{max} = the maximum expected binder strain, percent.

Meanwhile, the integrity parameter C which acts as the material integrity level can be calculated from Equation (8).

$$C = \frac{|G^*|\sin\delta_t}{|G^*|\sin\delta_{inital}} \tag{8}$$

where:

$|G^*|\sin\delta_t$ = the quotient of damaged value of $|G^*|\sin\delta$;
$|G^*|\sin\delta_{inital}$ = the initial undamaged value of $|G^*|\sin\delta$.

2.2.4. Bending Beam Rheometer Test

A BBR test was employed to characterize the low-temperature performance of SBS-MB, CR-MB and SBS/CR-MB before and after long-term aging according to ASTM D6648. The test temperature was −12, −18 and −24 °C and the average results of three replicates were used as the testing results. Using the interpolation method according to ASTM D7643-16 it is possible to determine the low service temperature (T_L) of asphalt binders from the BBR test in multiple testing temperatures. The critical temperature $T_{L,s}$ and ($T_{L,m}$) corresponding to stiffness = 300 MPa and m value = 0.3 were obtained by the regression Equations (9) to (10), respectively. The low service temperature (T_L) was defined in Equation (11).

$$\log_{10}(s) = a_1 + b_1 T \tag{9}$$

$$m = a_2 + b_2 T \tag{10}$$

$$T_L = \max(T_{L,s}, T_{L,m}) - 10 \tag{11}$$

where:

a_1, a_2, b_1 and b_2 are the regression coefficients;
T = the test temperature (°C);
$T_{L,s}$ = the critical temperature when S = 300 MPa (°C);
$T_{L,m}$ = the critical temperature when m = 0.3 (°C);
T_L = the low service temperature (°C).

3. Results and Discussion

3.1. Multiple Stress Creep Recovery Test Results

In the MSCR test, the major parameters used to identify the permanent deformation performance of the asphalt binder included non-recoverable compliance (J_{nr}), stress sensitivity ($J_{nr-diff}$) and percent recovery [22]. Table 2 shows the MSCR test final results for all bitumen binders. A detailed analysis of permanent deformation and deformation recovery performance of three kinds of modified binders before and after long-term aging is provided below.

Table 2. Multiple Stress Creep Recovery test results for three kinds of asphalt binder in different aging conditions.

Binder Type	Aging Conditions	%Recovery (0.1 kPa)	%Recovery (3.2 kPa)	R_{diff}, %	$J_{nr(0.1\ kPa)}$, 1/kPa	$J_{nr(3.2\ kPa)}$, 1/kPa	$J_{nr,diff}$, %
CRMB	Virgin	23.48	6.63	71.76	1.592	2.192	37.73
	TFOT	21.87	5.19	76.26	2.574	3.936	52.92
	PAV-20 h	17.03	4.13	75.74	3.863	6.542	69.43
SBSMB	Virgin	95.21	85.17	10.54	0.067	0.099	48.59
	TFOT	90.13	78.22	13.21	0.279	0.454	62.97
	PAV-20 h	82.11	52.18	36.45	0.443	0.733	65.45
	PAV-40 h	50.11	20.36	59.36	0.652	0.998	52.93
	PAV-80 h	35.48	10.29	71.00	0.913	1.384	51.57
SBS/CRMB	Virgin	96.01	89.16	7.14	0.031	0.042	37.37
	TFOT	92.71	82.71	10.78	0.132	0.178	34.97
	PAV-20 h	85.22	75.66	11.22	0.251	0.325	29.39
	PAV-40 h	81.21	63.29	22.06	0.294	0.402	36.71
	PAV-80 h	75.71	55.16	27.14	0.310	0.413	32.88

3.1.1. Analysis of Non-Recoverable Compliance at 3.2 kPa

Figure 1 displays the $J_{nr3.2}$ values for three kinds of modified bitumen at different long-term aging conditions. At the same testing temperature (70 °C), the $J_{nr3.2}$ values of CR modified bitumen were clearly larger than SBS and SBS/CR modified binders, indicating that the SBS modifier plays an important role in anti-permanent deformation. The addition of SBS and CR in SBS/CR modified bitumen further increased the rutting performance of binders due to the copolymer network as described in previous research. With the increased aging conditions the $J_{nr3.2}$ values of all kinds of bitumen increased, which was contrary to that of base asphalt due to aging leading to a stiffening effect for the binders. This phenomenon might be caused by the degradation of polymers during the aging process. Moreover, the $J_{nr3.2}$ values of SBS/CR modified binders increased slightly from the condition of virgin to the aging of PAV for 80 h, followed by SBS modified binders. These results illustrate that aging is detrimental to permanent deformation of the decomposition of the polymer caused by thermal oxygen aging. It must be noted that because of the poor performance of CR modified bitumen in the MSCR and LAS tests, the PAV aging test of CR modified bitumen for 40 h and 80 h was abandoned.

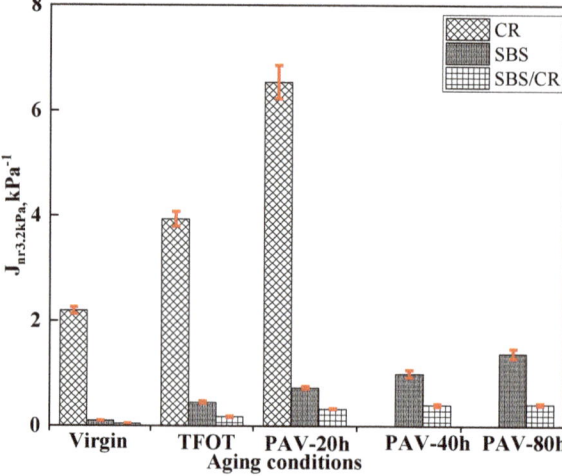

Figure 1. $J_{nr3.2}$ value for three kinds of modified bitumen at different aging conditions.

3.1.2. Evolution of Stress Sensitivity of All Kinds of Asphalt during the Aging Process

Figure 2 shows the $J_{nr,\,diff}$ values for three kinds of modified bitumen at different aging conditions. At the testing temperature of 70 °C, $J_{nr,\,diff}$ values of all tested binders were below 75%. Unlike the $J_{nr3.2}$ values of all tested binders, there was no significant difference found in $J_{nr,diff}$ values. Three kinds of modified binders held a similar stress sensitivity. Meanwhile, Jnr, diff values of three modified bitumen presented different trends during the process of aging. Jnr, diff values of CR and SBS modified bitumen before the PAV aging for 40 h increased with the extent of aging in contrast to that of SBS/CR modified bitumen. After PAV aging for 20 h, the Jnr, diff value at PAV aging for 40 h of SBS modified bitumen decreased suddenly, then decreased in the next aging condition of PAV aging at 80 h. However, the $J_{nr,\,diff}$ values of SBS/CR modified binders in different aging conditions was relatively stable. Lower $J_{nr,\,diff}$ values may be attributed to the aging of bitumen, which increased the binders' stiffness or a stronger cross-link network of polymers. Therefore, the $J_{nr,\,diff}$ values became complex during the process of aging [8].

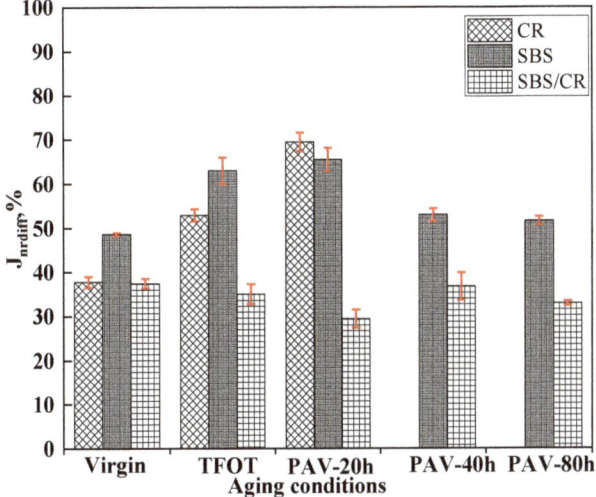

Figure 2. $J_{nr,diff}$ value for three kinds of asphalt at different aging conditions.

3.1.3. Analysis of Percent Recovery at 3.2 kPa

Percent recovery is a significant parameter which influences the deformation recovery of binder in the MSCR test. Figure 3 displays the percent recovery of the tested binders with the stress level of 3.2 kPa at 70 °C. CR modified bitumen owned the lowest percent recovery less than 10%, the other two modified bitumen possessed a much higher percent recovery. It indicated that the CR does not contribute to recovery behaviour of SBS/CR modified bitumen. The percent recovery of three modified bitumen decreased with the extension of aging time. Moreover, the percent recovery of SBS modified binders dropped rapidly after the aging of TFOT, but the percent recovery of SBS/CR modified binders decreased smoothly. This phenomenon may be attributed to the destruction of a three-dimensional network of SBS modifier in SBS modified bitumen, which decomposed faster than that in SBS/CR modified bitumen where there was a presence of carbon black, which released from CR during the process of the SBS/CR modified bitumen's preparation. The carbon black could protect the SBS molecule from oxidizing.

AASHTO M 332 put forward a method to detect the polymer in the bitumen, which is shown in Equation (12). Figure 4 shows the relationship between the percent recovery and the Jnr value at 3.2 kPa for three kinds of binders in different long-term aging conditions. Binders' percent recovery above the polymer modification curve demonstrates a good elastomeric behaviour. As showed in Figure 4, the percent recovery of CR modified bitumen is below the standard line in all aging conditions. Nevertheless, the SBS/CR- modified bitumen is always above the standard line regardless of aging conditions. The SBS modified binders were divided into two parts, on the one hand, when the PAV aging time was less than 40 h, the SBS-modified bitumen was up to the standard line and then below the standard line with increasing aging time. It demonstrated that the degradation of SBS modifiers in SBS/CR- modified binders was less than that of SBS modified binders. The SBS/CR modified bitumen employed an outstanding ability to resist aging.

$$R = \begin{cases} 29.37(J_{nr3.2})^{-0.2633}, J_{nr3.2} \geq 0.1 \\ 55, J_{nr3.2} < 0.1 \end{cases} \tag{12}$$

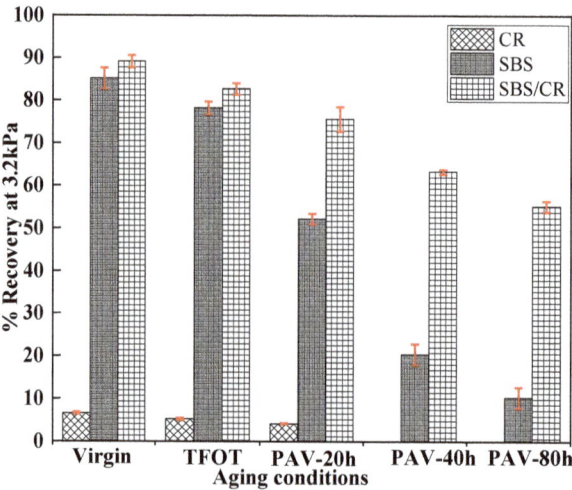

Figure 3. Percent recovery results for all kinds of bitumen at different aging conditions.

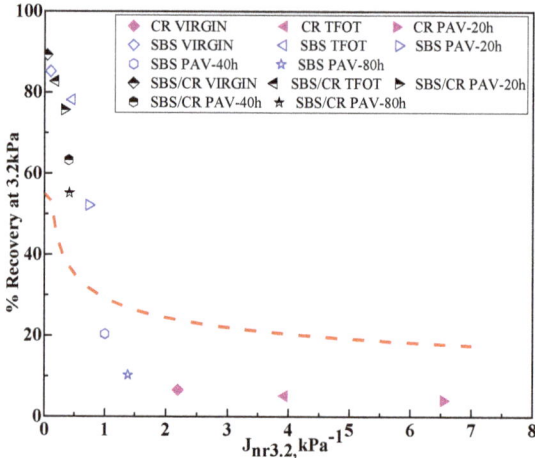

Figure 4. MSCR %R vs. J_{nr} at 3.2 kPa for all tested binders.

3.2. Linear Amplitude Scanning Test Results

3.2.1. Evolution of the Damage Intensity and Integrity Parameters of All Kinds of Asphalt during the Aging Process

Table 3 shows the LAS test fatigue damage parameter for CRMB, SBS MB and SBS/CR MB in different aging conditions. Figure 5 displays the relationship between damage intensity (D) and the integrity parameter (C) for all tested binders. A binder possesses better fatigue performance with lower values of C_1 and C_2 [23–25]. Table 3 shows that the value of great majority C_1 parameter increased with increasing aging time. However, the C_2 parameter was irregular with the aging time. Aged SBS/CR modified binders employed a lower C_1 value and higher C_2 parameter value than the virgin SBS/CR modified binder. The fatigue damage should be evaluated in combination with the relationship between the integrity parameter and damage intensity because of the overall effect of the decrease in integrity parameter, which depends on the combined effect of C1 and C2, as presented in Figure 5. Figure 5 shows that the integrity parameter was lost quicker while the thermal oxygen aging was further exacerbated. A significant difference of the loss rate in the integrity parameter with increasing aging conditions could also be found in the three kinds of modified bitumen. The loss rate of integrity parameter of all kinds of modified bitumen before the PAV aging employed a similar increased degree. However, when the aging time of PAV increased to 40 h, the loss rate of the integrity parameter of SBS modified binders was much larger than that of SBS/CR modified binders. The same results happened in the PAV aging for 80 h. The loss rate of the integrity parameter of the SBS/CR modified bitumen, which was aged in PAV for 80 h was nearly to that of the aged in PAV for 20 h. However, it was not be found in SBS-modified bitumen. Table 3 and Figure 5 illustrate that fatigue performance is affected intensely by the C_1 parameter.

Table 3. LAS test results of all tested binders based on viscoelastic continuum damage analysis.

Binder Type	Aging Conditions	C_1	C_2	A	B	α	τ_{max}
CR	Virgin	0.041	0.522	291329	2.474	1.237	0.142
	TFOT	0.042	0.530	388559	2.592	1.296	0.162
	PAV-20 h	0.047	0.515	530604	2.760	1.379	0.213
SBS	Virgin	0.050	0.473	3045026	2.894	1.447	0.222
	TFOT	0.047	0.496	1949513	2.888	1.444	0.242
	PAV-20 h	0.056	0.471	1627488	2.922	1.461	0.318
	PAV-40 h	0.058	0.484	657644	2.910	1.455	0.327
	PAV-80 h	0.061	0.536	641767	2.924	1.462	0.324
SBS/CR	Virgin	0.058	0.440	8017435	2.978	1.489	0.1988
	TFOT	0.043	0.504	4911318	3.026	1.513	0.2099
	PAV-20 h	0.042	0.513	3055835	3.042	1.521	0.2496
	PAV-40 h	0.044	0.505	3203419	3.05	1.525	0.2569
	PAV-80 h	0.047	0.498	3542636	3.07	1.535	0.2878

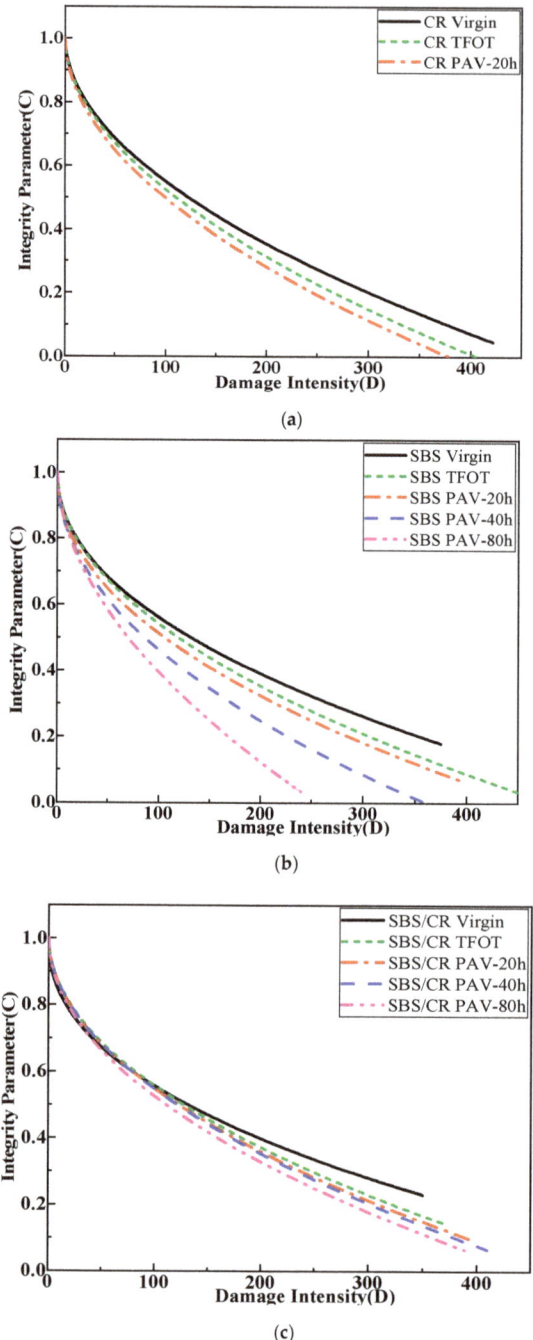

Figure 5. Effect of aging on the curve of integrity parameter vs. damage versus of Crumb Rubber modified bitumen (CRMB), SBS modified bitumen (SBSMB) and SBS/CRMB, (**a**) CR modified binders; (**b**) SBS modified binders; (**c**) SBS/CR composite modified binders.

3.2.2. Effect of Aging to Fatigue Life (N_f) during the Long-Term Aging Processes

The fatigue life (N_f) for all tested binders in the LAS test at strain levels of 1% and 10% are shown in Figure 6. As shown in Figure 6a, the opposite trend of the life cycles with the increasing of aging time at the 1% strain levels was observed in CRMB than that of SBSMB and SBS/CRMB. The value of N_f to fatigue failure of the CRMB became larger when the aging further happened compared with the results shown in Figure 5a. However, the same trend of the life cycles with increasing aging time at the 10% strain levels in all kinds of modified bitumen was consistent with the results shown in Figure 5a. This phenomenon may be led by the crumb rubber powder in CRMB, which was larger than modifiers in the other two modified bitumen. When the strain level was small, the crumb rubber power could strengthen the binders' fatigue life due to the elastic of crumb rubber. However, when the strain is larger, the crumb rubber power couldn't endure the deformation in the test, so the trend of CR modified binders' fatigue life with the increasing aging time became the same as the results shown in Figure 5a.

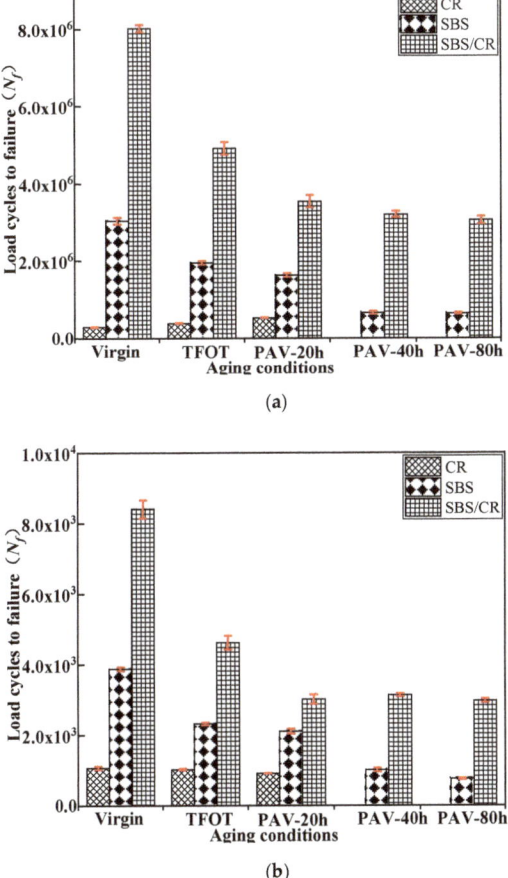

Figure 6. Effect of aging on LAS test fatigue life of CRMB, SBSMB and SBS/CRMB at different strain levels, (**a**) 1% strain level; (**b**) 10% strain level.

3.3. Low Temperature Performance

Figures 7 and 8 display the effect of aging on the stiffness modulus(S) and creep rate of CRMB, SBSMB and SBS/CRMB in different aging conditions. As shown in Figure 7, in contrast to the stiffness modulus of CRMB and SBS/CRMB, the stiffness modulus of SBS increased sharply after short-term aging, while the other two modified bitumen increased slightly at all test temperatures. The stiffness modulus of the three bitumen increased significantly after PAV aging for 20 h. With the PAV aging time increased, the stiffness modulus of the SBSMB increased faster than that of SBS/CRMB, which indicated that SBS/CRMB held a better ability for anti-aging. Figure 8 revealed that short-term aging had a great effect on the m value of all kinds of bitumen, which was consistent with the result of the stiffness modulus. Moreover, after long-term aging, the creep rate of all kinds of bitumen decreased. With the PAV aging time increased, the creep rate changed similarly to the stiffness modulus of the SBS and SBS/CRMB. However, some limitations still exist when evaluating the low temperature performance of bitumen by S and m values at a certain test temperature. In a number of studies it has been illustrated that it is important to establish a comprehensive indicator that combines m-value and stiffness modulus in the BBR test [15,26–28]. Table 4 shows that the SBSMB possessed the same low temperature performance grade (PG) at −24 °C in a low temperature in all aging conditions besides the virgin SBSMB. Therefore, the low service temperature and temperature difference ($\Delta T_L = T_{L,s} - T_{L,m}$) was calculated as shown in Table 4.

As shown in Figure 9, when the aging time increased, the PG low temperature decreased in all kinds of modified bitumen. CRMB and SBS/CRMB owned a lower PG low temperature than that of SBS modified bitumen. Moreover, the loss rate of PG low temperature in CR and SBS/CRMB was similar and smaller than that of SBSMB. It illustrated that CR plays an important role in low-temperature cracking of SBS/CRMB.

As shown in Figure 10, the temperature difference (ΔT_L) of all tested binders was smaller than 0, indicating the binder was more likely to break due to the lack of creep capacity (m-value controlled asphalt). The ΔT_L values of CRMB in the virgin condition was relatively closer to 0 °C, indicating that compared with the SBSMB and SBS/CRMB, the stiffness and the m-value of CR modified asphalt were more balanced. However, the ΔT_L values of CR modified asphalt after TFOT and PAV aging dropped sharply, revealing that the aging broke the balance between stiffness and the creep rate. However, for SBSMB, the ΔT_L values moved closer to zero during the process of aging. Meanwhile, the change of ΔT_L values of SBS/CR modified asphalt was erratic due to the aging conditions. These results illustrated that the influence of aging in ΔT_L values of SBSMB was in contrast to that of CRMB.

Table 4. PG Low temperature of BBR test for all tested binders.

Binder Type	Aging Conditions	$T_{L,m}$	$T_{L,S}$	T_L	$\Delta T_{L(S-m)}$
CR	Virgin	−22.77	−22.95	−32.77	−0.18
	TFOT	−21.63	−22.74	−31.63	−1.11
	PAV-20h	−19.96	−21.32	−29.96	−1.36
SBS	Virgin	−20.26	−23.06	−30.26	−3.02
	TFOT	−18.26	−20.46	−28.26	−2.20
	PAV-20h	−17.53	−19.56	−27.53	−2.03
	PAV-40h	−16.29	−18.16	−26.29	−1.87
	PAV-80h	−15.13	−15.53	−25.13	−0.40
SBS/CR	Virgin	−22.54	−23.58	−32.54	−1.04
	TFOT	−21.89	−23.46	−31.39	−1.57
	PAV-20 h	−20.39	−21.52	−30.39	−1.13
	PAV-40 h	−20.14	−20.93	−30.14	−0.59
	PAV-80 h	−18.58	−19.26	−28.58	−1.08

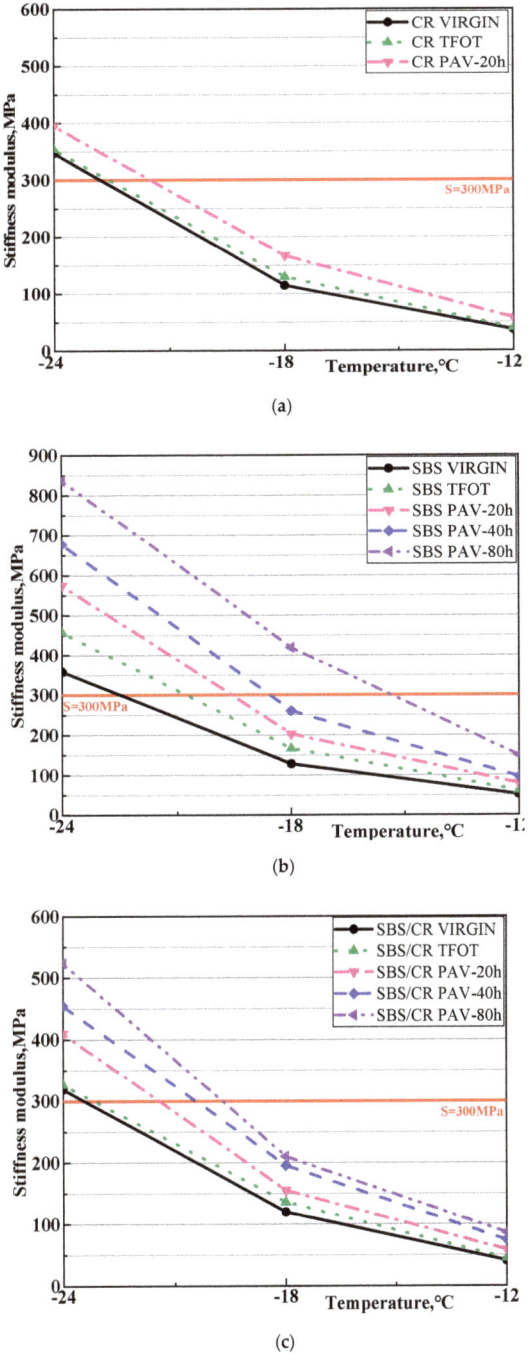

Figure 7. The evolution of stiffness modulus under different aging conditions of all selected asphalt binders, (**a**) Crumb Rubber modified binders; (**b**) Styrene-butadiene-styrene Modified binders; (**c**) Styrene-butadiene-styrene/Crumb Rubber Modified binders

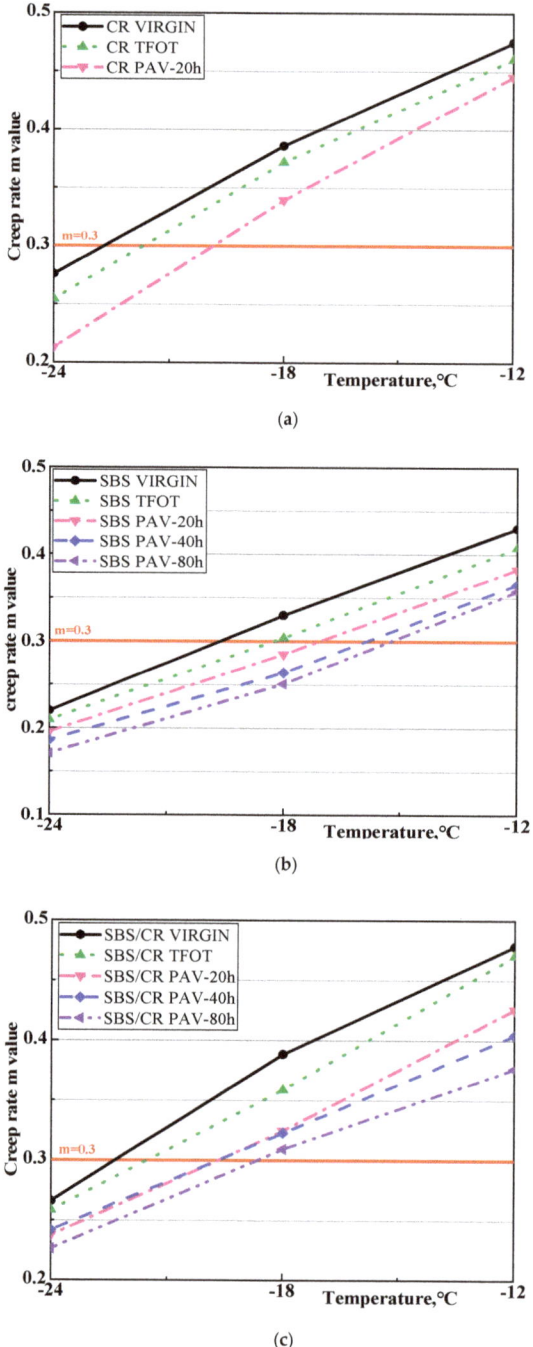

Figure 8. The evolution of creep rate under different aging conditions of all selected asphalt binders, (**a**) Crumb Rubber modified binders; (**b**) Styrene-butadiene-styrene Modified binders; (**c**) Styrene-butadiene-styrene/Crumb Rubber Modified binders.

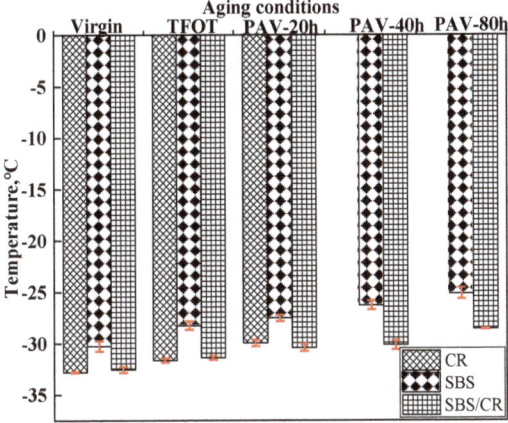

Figure 9. PG Low temperature of all tested binders.

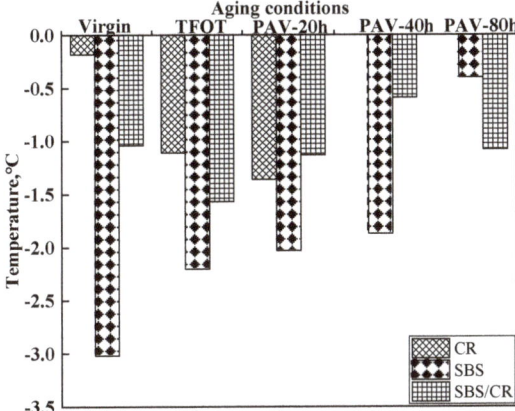

Figure 10. PG Low temperature difference of tested binders.

4. Conclusions

In this study, SBS/CRMB was prepared. Low temperature thermal and fatigue crack, permanent deformation performance of CR, SBS and SBS/CR modified asphalt in different aging conditions was also analyzed. The following conclusions were drawn:

(1) The SBS/CR composite modified asphalt possessed the best fatigue resistance, rutting resistance and a low temperature performance before and after different aging conditions. This showed the strong anti-aging ability of SBS/CRMB because of its flexibility and structure that remain in a good condition after long-term aging.

(2) Compared with CR and SBS modified asphalt in the virgin condition, the elastic property of CR in SBS/CRMB improved the ability to resist low temperature thermal and fatigue cracking at the range of low and middle temperatures. In the high temperature domain, the copolymer network greatly enhanced the elastic response of the asphalt SBS/CRMB, which shows better deformation recovery.

(3) Compared with CRMB and SBSMB under different long-term aging processes, there was a presence of carbon black, which released from the crumb rubber power during the process of

the SBS/CR modified bitumen's preparation. The carbon black could protect the SBS molecule from oxidizing.
(4) In contrast to CRMB and SBSMB, it is recommended that SBS/CRMB be used in the subgrade of a high speed railway. It is suggested that in future research, the properties of SBS/CR modified bitumen under different aging times of ultraviolet could be studied.

Author Contributions: The authors confirm contribution to the paper as follows: study conception and design: Y.Y., X.C., D.C.; data collection: G.X., X.C.; analysis and interpretation of results: L.L. and Y.S.; draft manuscript preparation: G.X. All authors reviewed the results and approved the final version of the manuscript.

Funding: This research was funded by National Natural Science Foundation of China grant number [No.51778136] and Technology Research and Development Project of China Railway (No. 2017G008-B). And The APC was funded by [No. 51778136].

Acknowledgments: Authors would like to thank the financial support from National Natural Science Foundation of China (No. 51778136) and Technology Research and Development Project of China Railway (No. 2017G008-B). The authors also wish to thank the reviewers for their valuable comments and suggestions concerning this manuscript.

Conflicts of Interest: The authors declare no conflict of interest.

References

1. Xue-Ning, M.A.; Liang, B.; Gao, F. Study on the Dynamic Properties of Slab Ballastless Track and Subgrade Structure on High-speed Railway. *J. China Railw. Soc.* **2011**, *33*, 72–78.
2. Yang, J. Research on Design of Waterproof and Drainage for Subgrade of Passenger Dedicated Railway. *J. Railw. Eng. Soc.* **2011**, *33*, 85–90.
3. Zhang, L.T.; Ke, W.U.; Zhang, L.W. Study on Waterproofing and Drainage System for Subgrade of Ballastless Track. *Soil Eng. Found.* **2010**, *31*, 19–24.
4. Zhao, G.; Liu, X.; Gao, L.; Cai, X. Characteristic Analysis of Track Irregularity in Subgrade Frost Heave Area of Harbin-Dalian High-speed Railway. *J. China Railw. Soc.* **2016**, *38*, 105–109.
5. Liu, S.; Yang, J.; Chen, X.; Yang, G.; Cai, D. Application of Mastic Asphalt Waterproofing Layer in High-Speed Railway Track in Cold Regions. *Appl. Sci.* **2018**, *8*, 667. [CrossRef]
6. Liu, S.; Yang, J.; Chen, X.; Wang, M.; Zhou, W. Design of Asphalt Waterproofing Layer for High-Speed Railway Subgrade: A Case Study in Heilongjiang Province, China. In *Transportation Research Board 96th Annual Meeting*; Transportation Research Board: Washington, DC, USA, 2017.
7. Liu, S.; Markine, V.L.; Chen, X.; Yang, J. Numerical Study on Application of Full Cross-Section Asphalt Waterproof Layer in CRTS III Slab Track. In *Transportation Research Board 97th Annual Meeting*; Transportation Research Board: Washington, DC, USA, 2018.
8. Zhou, Z.; Gu, X.Y.; Dong, Q.; Ni, F.J.; Jiang, Y.X. Rutting and fatigue cracking performance of SBS-RAP blended binders with a rejuvenator. *Constr. Build. Mater.* **2019**, *203*, 294–303. [CrossRef]
9. Liu, H.Y.; Hao, P.W.; Wang, H.N.; Adhikair, S. Effects of Physio-Chemical Factors on Asphalt Aging Behavior. *J. Mater. Civ. Eng.* **2014**, *26*, 190–197. [CrossRef]
10. Yin, F.; Arambula-Mercado, E.; Martin, A.E.; Newcomb, D.; Tran, N. Long-term ageing of asphalt mixtures. *Road Mater. Pavement Des.* **2017**, *18*, 2–27. [CrossRef]
11. Wang, P.; Dong, Z.J.; Tan, Y.Q.; Liu, Z.Y. Anti-ageing properties of styrene-butadiene-styrene copolymer-modified asphalt combined with multi-walled carbon nanotubes. *Road Mater. Pavement Des.* **2017**, *18*, 533–549. [CrossRef]
12. Chen, J.S.; Liao, M.C.; Shiah, M.S. Asphalt modified by styrene-butadiene-styrene triblock copolymer: Morphology and model. *J. Mater. Civ. Eng.* **2002**, *14*, 224–229. [CrossRef]
13. Yilmaz, M.; Yalcin, E. The effects of using different bitumen modifiers and hydrated lime together on the properties of hot mix asphalts. *Road Mater. Pavement Des.* **2016**, *17*, 499–511. [CrossRef]
14. Wu, S.P.; Pang, L.; Mo, L.T.; Chen, Y.C.; Zhu, G.J. Influence of aging on the evolution of structure, morphology and rheology of base and SBS modified bitumen. *Constr. Build. Mater.* **2009**, *23*, 1005–1010. [CrossRef]
15. Kok, B.V.; Yilmaz, M.; Geckil, A. Evaluation of Low-Temperature and Elastic Properties of Crumb Rubber- and SBS-Modified Bitumen and Mixtures. *J. Mater. Civ. Eng.* **2013**, *25*, 257–265. [CrossRef]

16. Shu, X.; Huang, B.S. Recycling of waste tire rubber in asphalt and portland cement concrete: An overview. *Constr. Build. Mater.* **2014**, *67*, 217–224. [CrossRef]
17. Abdullah, M.E.; Zamhari, K.A.; Buhari, R.; Nayan, M.N.; Hainin, M.R. Short Term and Long Term Aging Effects of Asphalt Binder Modified with Montmorillonite. In *Advanced Materials Engineering and Technology Ii*; Abdullah, M.M.A., Jamaludin, L., Abdullah, A., AbdRazak, R., Hussin, K., Eds.; Trans Tech Publications Ltd.: Zürich, Switzerland, 2014; Volume 594–595, pp. 996–1002.
18. Yu, X.; Leng, Z.; Wang, Y.; Lin, S.Y. Characterization of the effect of foaming water content on the performance of foamed crumb rubber modified asphalt. *Constr. Build. Mater.* **2014**, *67*, 279–284. [CrossRef]
19. Dong, F.Q.; Yu, X.; Liu, S.J.; Wei, J.M. Rheological behaviors and microstructure of SBS/CR composite modified hard asphalt. *Constr. Build. Mater.* **2016**, *115*, 285–293. [CrossRef]
20. Tan, Y.Q.; Guo, M.; Cao, L.P.; Zhang, L. Performance optimization of composite modified asphalt sealant based on rheological behavior. *Constr. Build. Mater.* **2013**, *47*, 799–805. [CrossRef]
21. Wang, Y.; Zhan, B.C.; Cheng, J. Study on preparation process of SBS/crumb rubber composite modified asphalt. In *Trends in Building Materials Research*; Zheng, J.J., Du, X.L., Yan, W., Li, Y., Zhang, J.W., Eds.; Trans Tech Publications Ltd.: Zürich, Switzerland, 2012; Volume 450–451, pp. 417–422.
22. Golalipour, A.; Bahia, H.U.; Tabatabaee, H.A. Critical Considerations toward Better Implementation of the Multiple Stress Creep and Recovery Test. *J. Mater. Civ. Eng.* **2017**, *29*, 7. [CrossRef]
23. Hintz, C.; Velasquez, R.; Johnson, C.; Bahia, H. Modification and Validation of Linear Amplitude Sweep Test for Binder Fatigue Specification. *Transp. Res. Rec.* **2011**, *2207*, 99–106. [CrossRef]
24. Hintz, C.; Bahia, H. Simplification of Linear Amplitude Sweep Test and Specification Parameter. *Transp. Res. Rec.* **2013**, *2307*, 10–16. [CrossRef]
25. Micaelo, R.; Pereira, A.; Quaresma, L.; Cidade, M.T. Fatigue resistance of asphalt binders: Assessment of the analysis methods in strain-controlled tests. *Constr. Build. Mater.* **2015**, *98*, 703–712. [CrossRef]
26. Liu, S.T.; Cao, W.D.; Fang, J.G.; Shang, S.J. Variance analysis and performance evaluation of different crumb rubber modified (CRM) asphalt. *Constr. Build. Mater.* **2009**, *23*, 2701–2708. [CrossRef]
27. Wang, H.N.; You, Z.P.; Mills-Beale, J.; Hao, P.W. Laboratory evaluation on high temperature viscosity and low temperature stiffness of asphalt binder with high percent scrap tire rubber. *Constr. Build. Mater.* **2012**, *26*, 583–590. [CrossRef]
28. Fu, Y.K.; Zhang, L.; Tan, Y.Q.; Meng, D.Y. *Low-Temperature Properties Evaluation Index of Rubber Asphalt*; Crc Press-Taylor & Francis Group: Boca Raton, FL, USA, 2016; p. 16.

© 2019 by the authors. Licensee MDPI, Basel, Switzerland. This article is an open access article distributed under the terms and conditions of the Creative Commons Attribution (CC BY) license (http://creativecommons.org/licenses/by/4.0/).

Article

Analysis of a Large Database of Concrete Core Tests with Emphasis on Within-Structure Variability

Angelo Masi, Andrea Digrisolo and Giuseppe Santarsiero *

School of Engineering, University of Basilicata, Viale dell'Ateneo Lucano, 10, 85100 Potenza, Italy; angelo.masi@unibas.it (A.M.); andrea.digrisolo@unibas.it (A.D.)
* Correspondence: giuseppe.santarsiero@unibas.it

Received: 23 May 2019; Accepted: 18 June 2019; Published: 20 June 2019

Abstract: In reinforced concrete (RC) structures, the compressive strength of concrete can play a crucial role in seismic performance and is usually difficult to estimate. Major seismic codes prescribe that concrete strength must be determined essentially from in situ and laboratory tests. Mean values obtained from such tests are the reference design values when assessing existing structures under seismic actions. The variability of concrete strength can also play an important role, generally requiring that various homogeneous domains are identified in a single structure, in each of which a specific mean value should be assumed as representative. This study analyzes the inter- and intra-variability of the concrete strength of existing buildings using a very large database made up of approximately 1600 core tests extracted from RC buildings located in the Basilicata region (Southern Italy). The analysis highlighted that concrete strength variability was dependent on the structures' dimensions as well as on the number of storeys. Moreover, the concrete strength of cores extracted from columns was found to be, on average, lower than that from beams, thus justifying the usual practice to extract cores mainly from columns, which results in a conservative approach as well as a more feasible one. Finally, some case studies were analyzed, specifically focusing on the effects of the within-storey variability. Conservative strength values, to be used especially in the case of vertical members subjected to high axial loads, are suggested.

Keywords: existing buildings; reinforced concrete; seismic vulnerability assessment; in situ concrete strength; variability of concrete strength

1. Introduction

In order to more effectively face natural risks, civil protection activities should be devoted to post-earthquake emergency, and above all, to prevention through a wide range of risk mitigation programs. Focusing in particular on the seismic risk reduction of buildings, vulnerability assessment is of fundamental importance for several reasons, among which: (i) to evaluate current safety conditions in view of possible urgent decisions; (ii) to define priorities and timescales in carrying out extensive strengthening programs; and (iii) to drive towards strengthening interventions that are highly effective in terms of the benefit–cost ratio.

The knowledge of concrete strength is essential to perform seismic and gravity load assessments of older reinforced concrete (RC) buildings. Reinforced concrete buildings, in fact, are frequently not provided with proper reinforcement detailing, thus brittle crises can occur in beam and column (or walls) members, and most importantly, in beam–column joints equipped with either deep beams [1] or wide beams [2,3]. The members' strength related to brittle failure modes is strongly related to concrete strength which, as a consequence, plays a key role in the assessment path.

Reliable seismic and gravity load assessments are also important in defining intervention priority, although it should be recognized that these decisions are based also on other socio-political demands and criteria.

Further, the knowledge of seismic vulnerability at the territorial scale can allow the elaboration of seismic scenarios for the most probable seismic events which can occur in a given area. Regarding this issue, some procedures aimed at large-scale vulnerability assessment of buildings (e.g., [4]) based on typological characteristics, have been developed. As these methods are based on poor data, their level of accuracy is limited.

In a similar context, other studies (e.g., [5]) evaluated the territorial distribution of seismic risk at the national scale, providing risk maps based on the current Italian hazard map [6]. The structural vulnerability was, in this case, evaluated with a detailed procedure (mechanical methods), based on the material properties reported in References [7,8], where mean and standard deviation values of the concrete strength of RC buildings built before 1970 were provided. The mechanical properties were derived from compression tests on concrete cubes at the Laboratory of Testing Materials of the University of Naples performed in the framework of the routine controls prescribed during the construction of new buildings. However, strength values obtained in such a way do not take into account the expected remarkable differences between current in situ strength (generally many years after the construction) and laboratory test results carried out during the construction.

For this reason, it appears appropriate to refer to the results of compression tests on concrete cores extracted from RC buildings at the time the seismic assessment is being carried out. Indeed, in Italy, starting from 2004, a huge (and still ongoing) assessment program was started to draw a picture of public buildings' vulnerability in order to allow owners and political authorities to make decisions and time scaling in view of an extensive strengthening plan. In fact, once the results of the seismic assessment are available, decisions specifically applicable to single buildings can be taken. First of all, the convenience of strengthening (either retrofitting or upgrading) with respect to replacing should be evaluated, and secondly, a comparison with other buildings should be made in order to establish priorities [9].

As a result of more than ten years of seismic assessments of public buildings, a large amount of data on concrete strength values obtained from a laboratory compression test on cores extracted from these buildings are available. As an example, Ferrini et al. [10] obtained data regarding approximately 1500 concrete cores extracted from 400 buildings located in the territory of the Tuscany region.

A similar database on concrete core tests (approximately 1600 specimens) is described and analysed in the present paper. They were collected as a result of the seismic assessments conducted in the Basilicata region on RC buildings whose construction period ranges between 1940 and 1990. Therefore, concrete strength distribution depending on the construction period can be determined. First, this can provide a useful tool for seismic vulnerability assessments at the regional level [11]. Secondly, the knowledge of mean strength values can be helpful, although not sufficient, in the structural evaluation of single buildings. In fact, in this case, the variability of concrete strength also matters, and it is quite important to identify homogenous areas over a single building, that is areas where a single value (generally the mean value according to seismic code provisions), can be assumed as representative. Some studies showed that the concrete strength variability inside a single building can be rather large, up to values of the coefficient of variation (CV) equal to 0.50 [12,13]. Other researchers showed that the CV values usually decrease when the strength increases [14].

In Reference [15], the high variability of the concrete strength, even along a single structural member, is discussed. The CV values ranging between 14% and 31% in different areas of the investigated member are found. Reasons for this variability and suggestions on how to manage it are provided in this paper. In Reference [16], a method to assess the variability of concrete strength along a single member using a rebound hammer test and compression tests on cores is proposed.

Accounting for the frequent large variability of concrete strength within a single building, a procedure for identifying homogenous areas is proposed in Reference [17], based on a first investigation phase made up of in situ, non-destructive tests. The results of these tests allow the definition of an effective, although as limited as possible, campaign of core drilling. Other studies provide general empirical expressions and approaches [18,19] estimating the in situ concrete strength variability using non-destructive test (NDT) results.

In the present paper, an updated version of the database reported in Reference [11] is studied and further analyses are made mainly aimed at providing indications on the within-building concrete strength variability, especially focusing on the influence of building size (particularly, the number of storeys) on CV values. Additionally, some remarks are also made by comparing strength values related to cores extracted from either beam or column members. Finally, some suggestions on the determination of the design strength to be used in the safety verifications prescribed by structural codes on existing buildings, under seismic loads, are proposed.

2. Description of the Core Tests' Database

Data analyzed in this study were collected within a large program of seismic vulnerability assessments of public buildings located in the Basilicata region (Southern Italy), which was managed by the local regional authorities (i.e., local governments), as prescribed by the national law OPCM 3274/2003 [20]. The assessment program involved strategic buildings, such as hospitals, and buildings at significant risk of collapse, such as schools. In this framework, a series of compression test results on concrete cores drilled from the buildings under evaluation was obtained. Therefore, the structural members (beam, column or wall) and the location where core specimens were extracted was the choice of the technician in charge of the specific evaluation. Results in terms of concrete compression strength were reported in the test certificates also providing other information regarding the core specimens (location in each structure, height, diameter, specific weight, etc.). The database is made up of 1572 concrete cores extracted from 346 structures, 280 of which are school buildings and 66 which are hospital buildings. Globally, 968 cores were extracted from schools (on average 3–4 cores/building) and 604 cores from hospitals (on average 9 cores/building).

In order to correct the strength values directly achieved on core specimens, possibly different from in situ strength, thus obtaining comparable results, the core strength values f_{core} were converted into the corresponding in situ strength f_c through the Relationship (1) reported in Reference [15].

$$f_c = (C_{H/D} \cdot C_{dia} \cdot C_{st} \cdot C_d) \cdot f_{core} \qquad (1)$$

where $C_{H/D}$ is the correction coefficient for the height/diameter ratio H/D different from 2.0; C_{dia} is the correction coefficient for the diameter of the core; C_{st} is the correction coefficient for the presence of reinforcing bars; and C_d is the correction coefficient for damage due to the drilling. Regarding this latter factor C_d, FEMA 274 [21] suggests assuming a constant value $C_d = 1.06$, while other researchers [22] suggest assuming 1.20 for $f_{core} < 20$ MPa, and 1.10 for $f_{core} > 20$ MPa, considering that the lower the strength, the higher the expected drilling damage. Following this approach, in the present paper, the values reported in Reference [23] have been used, assuming:

- $C_d = 1.30$ for $f_{core} \leq 10$ MPa;
- $C_d = 1.20$ if $10 < f_{core} \leq 20$ MPa;
- $C_d = 1.10$ if $20 < f_{core} \leq 30$ MPa;
- $C_d = 1.00$ if $f_{core} > 30$ MPa.

In the following, some basic analyses on the database in terms of f_c values are reported.

Data was first analyzed in respect to the construction period of buildings from which cores were extracted. Four construction periods, being related to the enforcement of important structural codes on an RC buildings in Italy, were considered, namely, <1961, 1961–1971, 1972–1981, and >1981.

Table 1 reports the number of cores and buildings for each construction period with the related mean and median values of in situ strength fc, as well as the related dispersion in terms of standard deviation and coefficient of variation CV.

Table 1. Main statistical values of the available data on concrete strength f_c.

Construction Period	<1961	1961–1971	1972–1981	>1981	ND [1]
Number of buildings	23	112	114	77	20
Number of cores	111	568	578	250	65
Mean value (MPa)	14.96	20.08	22.31	27.37	23.47
Median value (MPa)	14.10	18.83	20.81	25.32	22.77
Standard deviation (MPa)	6.58	8.26	11.11	10.12	7.81
Coefficient of variation	0.44	0.41	0.50	0.37	0.33

[1] ND: not defined.

Data reported in Table 1 are also displayed in the graphs in Figure 1. As can be seen, most of the buildings, and consequently the cores, belong to the period 1961–1981, i.e., the years of the maximum economic growth in Italy and also in the Basilicata region. As expected, both the mean and median values of concrete strength increased with time, possibly due to the enforcement of building codes providing more severe rules on the control of materials' quality during the construction of new buildings.

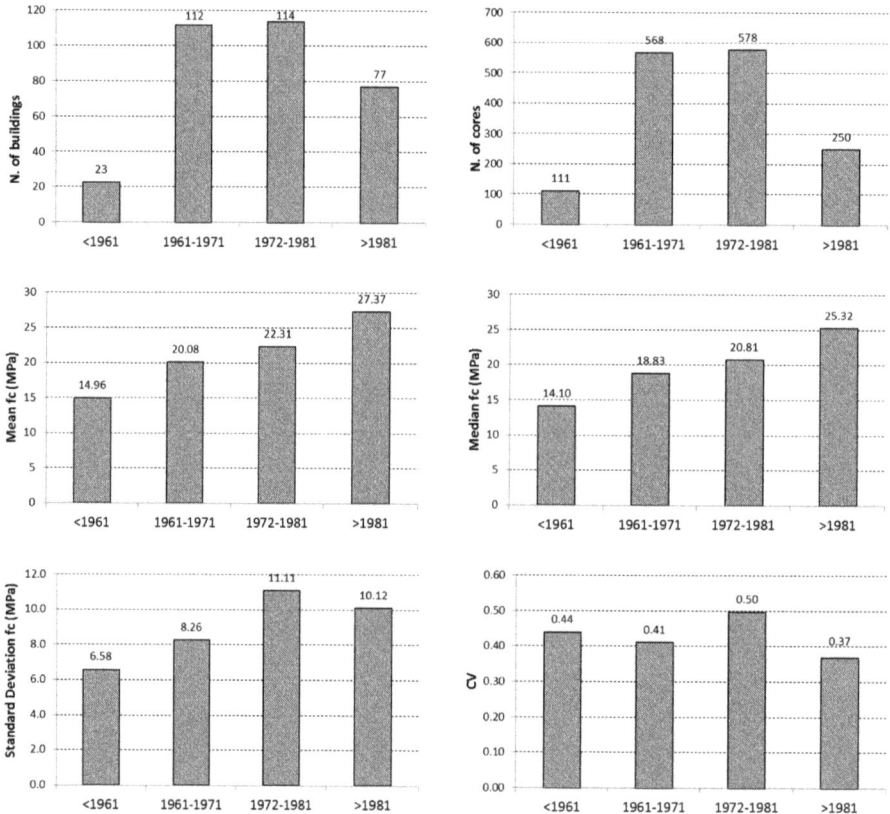

Figure 1. Distribution of the main statistical data related to the core strength values as a function of the construction period.

Vice versa, the standard deviation and coefficient of variation values did not show a well-defined trend over time. However, as can be noted, the dispersion was generally high with the CV values being in the range of 0.37–0.50.

In Figure 2, the distributions of f_c values in the four construction periods are displayed, confirming the high dispersion, especially for the period 1972–1981. It is worth noting that also for the more recent periods (i.e., 1972–1981 and >1981) there was a significant quota of buildings with very low concrete strength (<10 MPa). Moreover, the last two periods did not show an improvement in terms of concrete homogeneity, as could be expected after the enforcement of a new code on RC constructions.

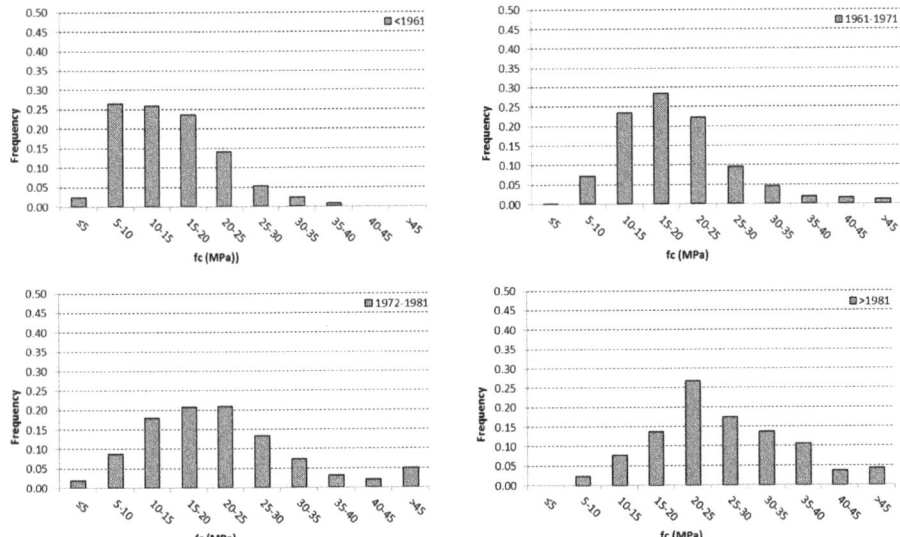

Figure 2. Distribution of the f_c values in the construction periods under consideration.

Figure 3 displays the variability of the mean concrete strength as a function of the number of storeys. For each group of buildings with a given number of storeys, the mean concrete strength was computed, discarding buildings with fewer than five cores. As a result, the database is made up of 100 building. As can be seen, concrete strength had a clear trend of increasing with the number of storeys. It can also be noted that an evident difference was visible between buildings with four storeys or less and buildings with five to eight storeys. This could be due to the better concrete quality, workmanship, and supervision adopted for the construction of larger buildings.

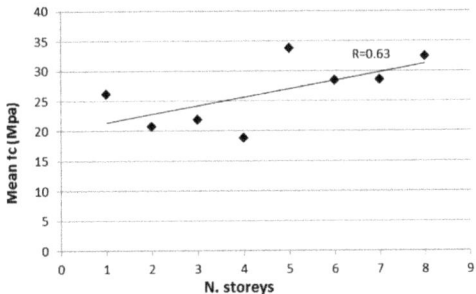

Figure 3. Mean concrete strength as a function of the number of storeys.

3. Analysis of Strength Values by Structural Member Type

When planning a destructive tests campaign, a crucial issue relates to the choice of the sampling points. Current seismic codes on the evaluation of RC buildings provide rules forcing the extraction of cores from both column and beam members. However, due to the practical constraints, difficulties in extracting cores from beams are well known among structural engineers, and more importantly, in some cases, extraction can be almost impossible (e.g., flat or wide beams having the same thickness as the adjacent floor slab). Moreover, even in the case of deep beams, firmly fixing the drilling machine could be rather difficult. As a result, in most cases, core drilling is carried out on vertical members (column and walls), and only in a few cases on beams.

This is clearly revealed by the database under study, where most cores were extracted from columns, that is approximately 1400 out of a total of ~1600 cores.

In order to understand if/how the strength of the cores drilled from columns can be assumed as representative also of beams and then used to determine the design value to be used in safety verifications, the database was analyzed to highlight possible differences between column and beam cores (Table 2). The comparison was made considering only the buildings where cores extracted from both columns and beams were available and assuming that the concrete used in the column members had the same mix design as that used in the beams, as it is a common construction practice. Therefore, the total number of values to be compared decreased to 240 (156 cores from columns and 73 from beams, extracted from 41 buildings). The comparison is displayed in Figure 4 as a function of different construction periods, where the period <1961 was not considered due to the low number of available values. Note that the mean values reported in Figure 4 were computed by averaging, in each period, the mean values of core strength (separately for columns and beams) related to each single building.

Table 2. Statistical data of the subset of concrete strength values regarding buildings with cores extracted from beams and columns (B = beams, C = columns).

Construction Period	<1961	1961–1971	1972–1981	>1981
Number of buildings	2	15	11	13
Number of cores (B)	8	31	23	22
Number of cores (C)	16	56	44	40
Mean value (MPa) (B)	23.8	19.9	24.7	26.6
Mean value (MPa) (C)	17.2	18.0	19.6	22.7
Standard deviation (MPa) (B)	7.20	5.89	9.88	5.49
Standard deviation (MPa) (C)	2.61	5.38	7.80	4.69
Coefficient of variation (B)	0.30	0.30	0.40	0.21
Coefficient of variation (C)	0.15	0.30	0.40	0.21

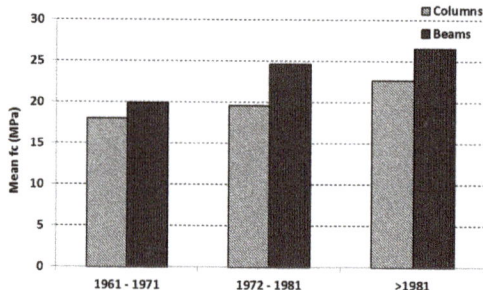

Figure 4. Mean strength values of cores extracted from columns and beams (computed by averaging, in each period, the mean values of the core strength from columns and beams relevant to each single building).

As displayed in Figure 4, for all construction periods, the mean strength of cores extracted from beams was always higher than that from columns, with a difference in the range 1.10–1.25.

In order to evaluate whether the observed differences were significant, Students' *t*-tests were performed on the two strength value sets (columns versus beams). The tests can be applied in two different ways:

- two-tailed test: to know whether the mean values are equal;
- one-tailed test: to know whether a mean is higher than the other.

If one cannot exclude that a mean value is higher than the other, the two-tailed test is preferable.

The *t*-test assumes a null hypothesis (H_0), meaning that the mean related to the two sets values (μ_C for columns, μ_B for beams) belong to the same population, the differences being due to the random variations, and the alternative hypothesis (H_1) that the two mean values belong to different populations, that is:

$$H_0: \mu_C = \mu_B \tag{2}$$

$$H_1: \mu_C \neq \mu_B \tag{3}$$

The test was carried out referring to two significance levels: $\alpha = 0.05$ and $\alpha = 0.01$.

Table 3 reports the statistics result T and the limit values for the two significance levels: T0.05 and T0.01. In the last two rows, the result of the *t*-test for each construction period is reported, where OK means that the hypothesis H_0 cannot be rejected.

Table 3. The *t*-test on concrete strength values extracted from vertical members and beams.

Construction Period	1961–1971	1972–1981	>1981
T	−1.53	−2.29	−2.90
$T_{0.05}$	2.28	2.29	2.30
$T_{0.01}$	2.88	2.91	2.91
$R_{0.05}$	OK	OK	NO
$R_{0.01}$	OK	OK	OK

For a significance level $\alpha = 0.05$, the *t*-test yielded a positive result for two out of the three construction periods, while for a significance level $\alpha = 0.01$, a positive result was found for all the construction periods. Therefore, the differences between the two populations were negligible, and consequently, the strength values of columns and beams can be considered as belonging to the same population.

Further, there were two additional aspects to be addressed in order to perform core drilling on column members: i) core drilling from columns is generally less expensive and time-consuming; ii) columns play a more important role in the load-bearing structural system under both vertical and seismic loads.

In sum, although the above findings should be considered valid with respect to the database under study, they are nevertheless important as performing core drilling only on columns appears to be on the safe side, with their concrete strength being, on average, lower than that of beams, although they belong to the same population. Therefore, the strength values found on columns' cores can also be representative of beams' strength. This remark is consistent with the work in progress to update the Eurocode 8 (EC8)-Part 3: Assessment and retrofitting of buildings and bridges [24], where the recommended minimum requirements for different levels of testing no longer prescribe that cores need to be extracted from each type of primary element (e.g., column and beam), as is prescribed in the current version of EC8-3. Moreover, this also affects the shear strength of beam–column joints since they are cast along with beams. Due to this, they have to be assumed to have the same concrete strength.

4. Analysis of Within-building Variability

Previous analyses were devoted to outlining the general features of the whole database under consideration. However, the concrete strength properties can show high variability in the same building, between different storeys, and even at the same storey, although it can be assumed that the concrete used at different storeys has the same mix design, as it is in usual construction practice.

In order to analyze the within-building variability, further analyses were performed only on buildings where a large number of cores were available. Assuming that this number is not less than 5, the new database includes 100 buildings and 802 cores (Table 4).

Table 4. Statistical data of the subset of concrete strength values regarding buildings with at least five cores.

Construction Period	<1961	1961–1971	1972–1981	>1981	ND
Number of buildings	8	33	40	16	3
Number of cores	69	257	339	105	32
Mean value (Mpa)	17.45	20.78	22.51	30.18	21.60
Median value (Mpa)	18.40	18.65	20.39	29.43	18.45
Standard deviation (Mpa)	4.46	7.12	9.92	8.06	5.85
Coefficient of variation	0.26	0.34	0.44	0.27	0.27

Relating the coefficient of variation to the mean value of the concrete strength in each building (Figure 5), no correlation was found, as already reported in other studies [14,25]. This reveals that an increase in the average quality of concrete in existing buildings is generally not accompanied by a correspondent decrease in the concrete strength variability.

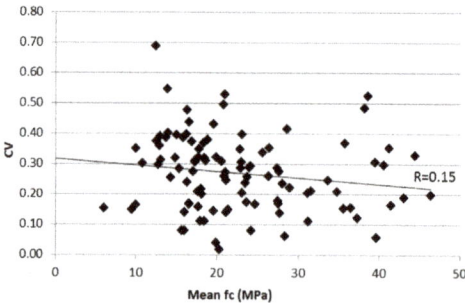

Figure 5. Comparison of CV and mean values computed in each building.

The estimation of concrete strength for design and assessment purposes is related to the choice of the knowledge level (KL) of the structure under study. Three KLs are defined (limited, normal, and full knowledge) in order to choose the appropriate confidence factor (CF) value to be adopted to reduce the concrete strength in the evaluation process. Once a knowledge level is assumed (KL1, KL2 or KL3), the current European and Italian seismic codes define the number of cores to be extracted depending on the total storey area, implicitly assuming that a single storey can be considered a homogeneous area. This assumption appears justifiable given that, except in the case of a very large floor area, the members of a single storey are generally cast almost simultaneously, thus similar concrete properties should be expected. On the other hand, different storeys can be cast after a period of time, leading to possible differences, for example, due to the modified environmental curing conditions (temperature, moisture). This supports the preliminary assumption that different storeys might represent different homogeneous areas.

On the basis of the above remarks, it is expected that the global dispersion within a whole building should increase with the number of storeys. In fact, Figure 6b displays that, in general, the percentage of buildings where the CV value is higher than 15% (a typical reference value given in the literature and in some structural codes) increases with the number of storeys, with the only exception being the five-storey group, where two buildings out of six have CV < 15%.

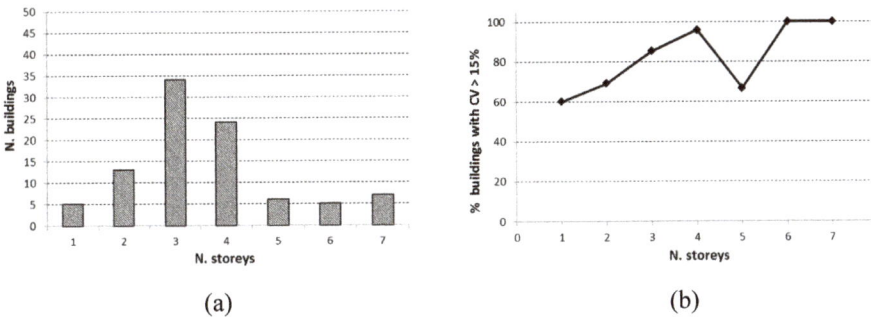

Figure 6. (**a**) Distribution of buildings by number of storeys; (**b**) number of buildings with CV higher than 15%.

Further, the analysis of data showed that the average floor area did not vary significantly with the number of storeys, so the variation of CV values, shown in Figure 6b, was essentially determined by the number of storeys.

This result confirms that considering a whole building as a single homogenous domain can be inappropriate in most cases, as the assumed mean strength value would be associated with an unacceptable high variability.

As can be seen in Figure 7, increasing the number of cores extracted from each storey in a building generally did not improve the achieved results in terms of a decrease in dispersion (i.e., CV value does not decrease), similar to the findings in Reference [26], and in contrast to the approach of FEMA 356 [27], which suggests increasing the number of cores in order to reduce the variability. This confirms that different storeys are likely to be homogeneous areas different from each other, and moreover, the variability of concrete strength is often intrinsic to RC buildings.

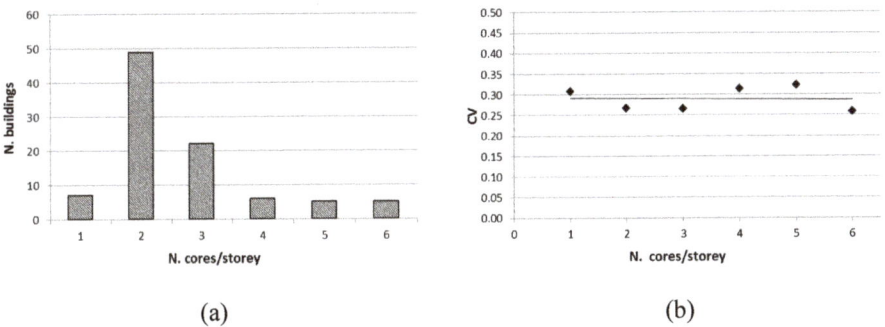

Figure 7. (**a**) Distribution of buildings as a function of the core number extracted from each storey; (**b**) CV values in relation to the mean number of cores extracted by single storeys.

Looking at the graph in Figure 8, a general increase in the dispersion can be observed when the number of storeys increases. In fact, the mean value of CV is higher for taller buildings, as demonstrated by the trend of the mean values of CV (dotted line in Figure 8). The median value had the same trend, while the standard deviation appeared to be independent of the number of storeys. Indeed, the results showed that even for one-storey buildings, CV was higher than 15%, being around 20%. This result suggests that one-storey buildings could also be made of more than one homogenous area due to the size of the building in plan. In fact, when the size increases, the concrete is likely to belong to more than one batch.

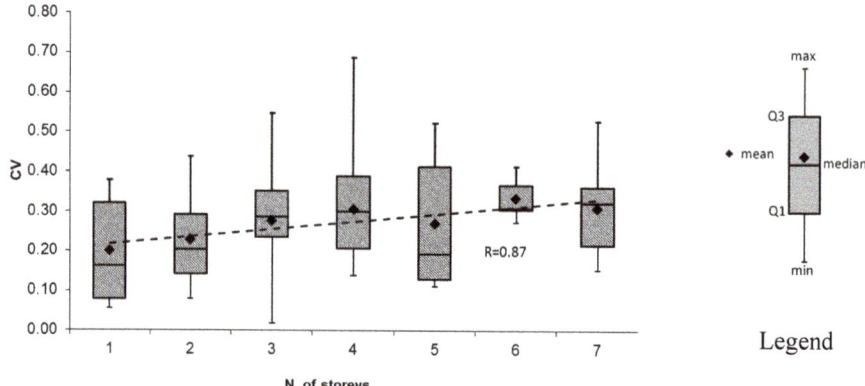

Figure 8. Box-plot of statistical data with respect to the number of storeys.

The mean CV value correlates well with the number of storeys, as can be seen in Figure 9, where a high correlation factor ($R = 0.91$) was found using the following logarithmic regression relationship:

$$CV = 0.0628 \ln(n_p) + 0.1979 \quad (4)$$

where n_p is the number of storeys.

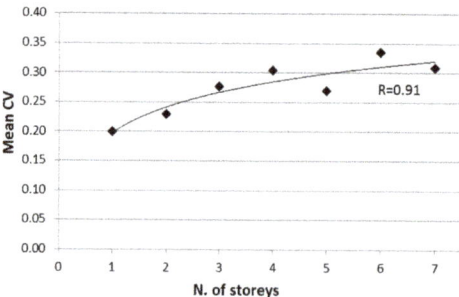

Figure 9. Logarithmic regression between mean values of CV and number of storeys.

This finding confirms the importance of appropriately identifying the different homogeneous areas in each building accounting for the fact that, in general, the variability of concrete properties increases with the building dimensions, especially with the number of storeys. Further, this finding suggests taking into account the average expected dispersion of concrete strength as a function of the number of storeys when making vulnerability assessments at the territorial scale (i.e., when the seismic vulnerability needs to be assigned to a single or a class of buildings, necessarily assuming a single strength value for each building).

Analysis of Within-Storey Variability

To further develop the analyses carried out in the previous sections by examining also the within-storey variability, five buildings with a number of storeys ranging from 3 to 7 and with at least 4 cores per storey, were selected (Table 5).

Table 5. Selected buildings for the analysis of within-storey variability.

Building ID	Construction Period	No. of Storeys	No. of Storeys with CV < 0.15	No. of Cores (Total)	No. of Core/Storey	No. of Core/Storey (min–max)	CV (Total)[1]	CV Storey (min–max)
1	1961–1971	6	1/6	33	5.5	5–6	0.24	0.10–0.30
2	1961–1971	7	1/7	63	9.0	6–12	0.25	0.13–0.32
3	<1961	5	0/5	34	6.8	6–8	0.27	0.20–0.31
4	1972–1981	6	0/6	56	9.3	6–13	0.25	0.16–0.35
5	1961–1971	3	0/3	17	5.7	5–6	0.34	0.16–0.39
	Total	27	2/27	203	7.5			

[1] referred to whole building.

Figure 10 shows the distribution of the CV values relevant to the whole building (black circles) and to each storey (grey diamond-shaped dots). Firstly, there is a high variability of concrete strength, with the CV values being higher than 0.15 most of the time (although at least four cores per storey were extracted), as shown in the fourth column of Table 5 that reports the number of storeys with a CV < 0.15 out of the total number of storeys for each building. Moreover, it is worth noting that the CV value computed for the whole building was quite different from the CV values relevant to each storey. Once again, this confirms the need to identify homogenous areas in building structures in order to obtain mean values of concrete strength that are adequately representative of each area. On this issue, a broad discussion with remarks and suggestions on a possible approach to be followed is reported in Reference [17].

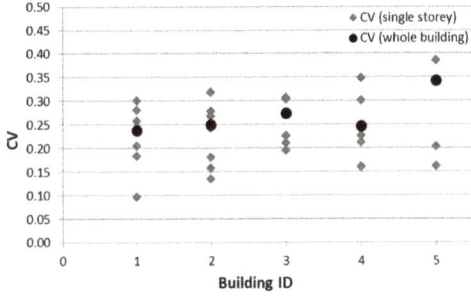

Figure 10. CV values computed in the single storeys and in the whole building.

Previous analyses demonstrated the high variability of concrete strength in a single building and even within a single storey, in most cases, is due to the intrinsic variability of the material properties. Increasing the number of cores did not reduce the dispersion of strength values. This latter result was evidenced by the high number of buildings having a CV larger than 0.15. One should derive that a single strength value cannot be representative of more than one storey, which represents the maximum extent of the so-called homogenous areas. Indeed, in some cases, more homogenous areas should be set for a single storey, so that any possible irregularity in the plan due to the concrete strength variability [28,29] is not neglected. Moreover, also in the case of seismic evaluation, dealing with high within-storey variability using the mean value could be not on the safe side due to the different role of concrete strength in determining the capacity of structural members (e.g., ductile versus fragile members).

A lower strength value on the larger structural members (especially in the case of columns) can lead to remarkable negative effects on the total structural capacity. This is especially true for frame structures with high axial load values on the columns. In order to clarify this aspect, a simple example was considered: a shear-type portal frame made up of two columns with dimensions of 300 × 300 mm for column n.1 and 300 × 600 for column n.2 (see Figure 11), provided with an amount of longitudinal reinforcement of 0.5% (typical of pre-code buildings), meaning $A_{s1} = 450$ mm² for column n.1 (four

12 mm diameter bars) and $A_{s2} = 900$ mm² for column 2 (six 14 mm diameter bars). The storey height was set to h = 3000 mm.

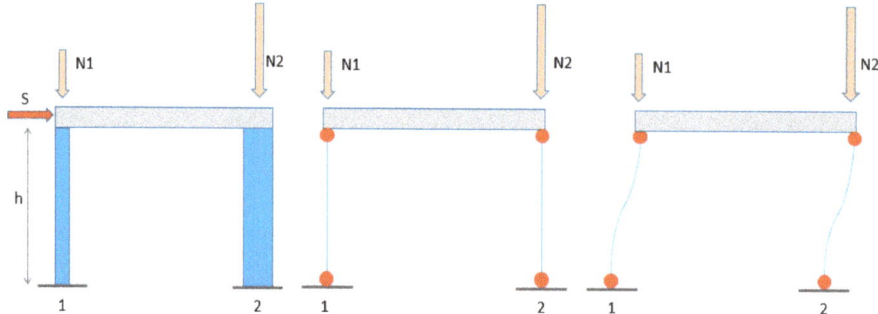

Figure 11. Example portal frame.

As for concrete strength, the characteristics of the 2nd storey of building no.1 (Table 5) were considered, provided with a mean concrete strength $f_{med} = 19.53$ MPa and minimum strength $f_{min} = 10.68$ MPa. The steel yielding stress was assumed as $f_y = 320$ MPa.

Axial load values proportional to the columns' gross area were assumed (as in typical design practice), i.e., equal to 40% of the ultimate axial load determined as $N_u = A_c \cdot f_{min}$. This means that the axial load value on column 1 was about $N_1 = 360$ kN, while that on column 2 was $N_2 = 720$ kN.

Based on the assumption of shear type, a double plastic hinge is expected to develop at column ends in a storey failure mechanism due to the horizontal actions. So, the frame base shear is directly proportional to the yielding moments developed at column ends. The total horizontal strength capacity of the structure is then:

$$S = \frac{2 \cdot M_{y1} + 2 \cdot M_{y2}}{h} \quad (5)$$

Yielding moment values for the two columns using either f_{med} (case "a") or f_{min} (case "b") can be computed:

(a) $M_{y1a} = 60.0$ kNm, $M_{y2a} = 238.2$ kNm
(b) $M_{y1b} = 50.6$ kNm, $M_{y2b} = 185.6$ kNm

Now, two scenarios are assumed:

1. Column 1 having concrete strength f_{med} and column 2 f_{min}
2. Column 1 having concrete strength f_{min} and column 2 f_{med}

In scenario 1, the total strength capacity is:

$$S_1 = \frac{2 \cdot M_{y1a} + 2 \cdot M_{y2b}}{h} = 163.7 \text{ kN} \quad (6)$$

While in scenario 2:

$$S_2 = \frac{2 \cdot M_{y1b} + 2 \cdot M_{y2a}}{h} = 192.5 \text{ kN} \quad (7)$$

As can be seen, scenario 1 results in a total base shear 15% lower than scenario 2.

This means that assuming a strength value equal to the mean concrete strength can lead to overestimation of the storey shear capacity due to the possible presence of low strength values on structural members such as larger columns or walls. This happens even for ductile mechanisms such as flexure for columns, for which the concrete strength plays an important role in the presence of significant axial load values. In order to visualize this situation, the base shear values of the example portal frame in scenarios 1 and 2 were plotted as a function of the axial load.

As can be seen from Figure 12, resulting base shear values of the portal with inverted concrete strength values (scenarios 1 and 2) diverge when axial load increases. In this regard, it should be taken into account that existing RC buildings, designed only according to gravity loads, often have high axial load values as a result of the non-seismic design approach.

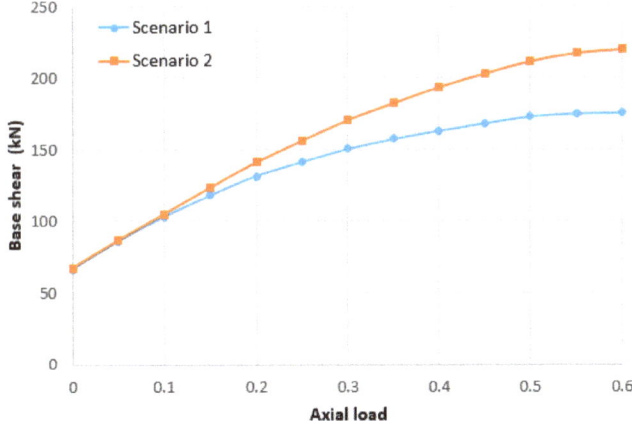

Figure 12. Base shear as a function of the normalized axial load.

Therefore, for storeys with high concrete strength variability (CV > 0.15), it is advisable to use the mean value minus one standard deviation, as proposed by FEMA 356 [27], in order to obtain more conservative results. However, it should be noted that seismic codes frequently allow the collection of an insufficient number of cores to determine CV and standard deviation. In fact, for small–to medium-sized residential buildings, at most, three cores per storey can be drilled (KL3), irrespective of the floor surface.

Therefore, it is strongly suggested to integrate destructive tests with non-destructive tests in order to obtain a sufficient number of strength values [17]. However, in situ material tests give rise to remarkable direct costs (drilling and non-destructive testing operations with related repair works) and indirect costs (related to occupancy disruption). Due to these issues, concrete strength values alternative to the mean minus one standard deviation should be assumed in cases where few strength values are available.

In order to derive some indication, the distributions of concrete strength along the height of the five selected buildings are plotted in Figure 13. In particular, for each storey, the mean value f_{med}, the minimum f_{min}, and $f_{med-\sigma}$ are displayed.

As can be seen from Figure 13, $f_{med-\sigma}$ is generally very close to f_{min}. For example, in buildings no. 1, 2, and 4, the two strength values had almost the same trend, except for the 1st storey of buildings no. 2 and 4 and the 2nd storey of building no. 1. Buildings no. 3 and 5 showed larger differences at the first storeys, although a good correspondence between f_{min} and $f_{med-\sigma}$ was generally found.

Based on the above considerations, in the case of seismic assessment where concrete strength was based on few cores per floor, and high scatter was observed among related strength values; it seems advisable to use the minimum concrete strength value of each floor in place of the mean. In fact, the results showed that the minimum value was somehow representative of $f_{med-\sigma}$, that is, a more conservative strength value is suggested by FEMA356 [27] when dispersion is high.

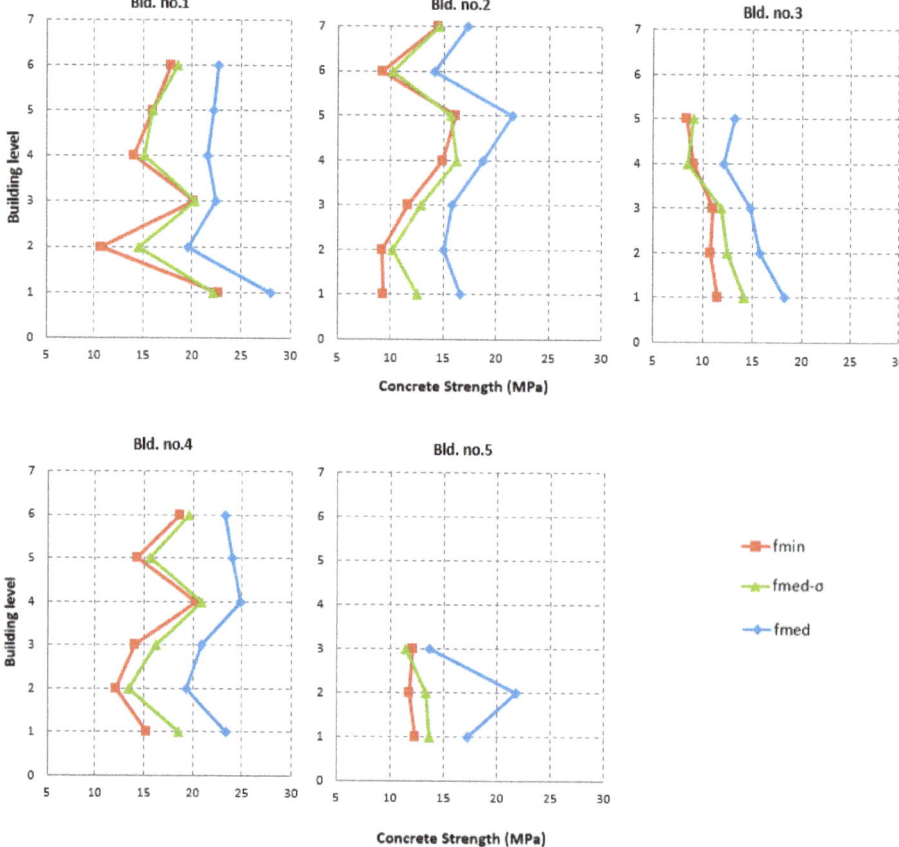

Figure 13. Comparison of the strength values at different heights for the five selected buildings.

5. Concluding Remarks

The availability of a large database of seismic assessments of RC buildings located in Basilicata (Italy) allowed the analysis of concrete strength properties (evaluated by means of core tests), in terms of mean values and variability, across different construction periods. First, the analyses showed that the concrete strength in column members was, on average, lower than that in beams. This outcome has important implications in practical seismic assessments of existing RC buildings, justifying the extraction of concrete cores only from columns, being both on the safe side and more feasible. Further, determining the concrete strength directly on columns appears still more remarkable accounting for their crucial role in carrying gravity loads.

After these preliminary analyses, the within-building variability of concrete strength was investigated. It was found that the coefficient of variation of concrete strength was mostly higher than the limit value provided by structural codes (e.g., Italian code), and was almost independent of the number of cores extracted from each storey. Looking at large-scale vulnerability assessments, where mean values as well as variability of concrete strength are required, a logarithmic relationship between CV and the number of storeys was found.

The variability of concrete strength was specifically analyzed in five buildings where a large number of cores was available (i.e., at least four cores per storey). The results showed that the variability across the whole building was comparable to that relevant to each single storey.

Analyzing the within-storey variability of concrete strength, it was found that using the mean value in the presence of a high dispersion of strength values can lead to an overestimation of the storey capacity, especially when lower strength values were related to the larger structural member such as large columns or shear walls. The influence of axial load values in this regard was also examined, underlining that this also happens in the presence of ductile mechanisms, such as flexure in column members. In order to overcome this, it is suggested to assume as a design value the mean value minus one standard deviation, provided that a sufficient number of test values are available for each floor. Taking into account that in usual practice, due to the costs and occupancy disruption constraints, the number of in situ tests at each storey is generally less than three, and it is not possible to calculate dispersion parameters (CV and standard deviation). In these cases, particularly in the case of significant differences among the available strength values, it is suggested to assume the minimum value from tests as the design value.

Author Contributions: A.M. conceived the study and the methodology and contributed to write the paper; A.D. collected and analyzed the data and contributed to write the paper; G.S. contributed to the analysis and interpretation of data and to write the paper.

Acknowledgments: The work reported in this paper was partly carried out within the framework of the DPC-ReLUIS 2019-21 Project, Work Package 4 "Seismic Risk Maps".

Conflicts of Interest: The authors declare no conflict of interest. The founding sponsors had no role in the design of the study; in the collection, analyses, or interpretation of data; in the writing of the manuscript, and in the decision to publish the results.

References

1. Masi, A.; Santarsiero, G.; Nigro, D. Cyclic tests on external RC beam-column joints: Role of seismic design level and axial load value on the ultimate capacity. *J. Earthq. Eng.* **2013**, *17*, 110–136. [CrossRef]
2. Masi, A.; Santarsiero, G. Seismic tests on RC building exterior joints with wide beams. *Adv. Mater. Res.* **2013**, *787*, 771–777. [CrossRef]
3. Santarsiero, G.; Masi, A. Key mechanisms of the seismic behaviour of external RC wide beam–column joints. *Open Construct. Build. Technol. J.* **2019**, *13*, 36–51. [CrossRef]
4. Zuccaro, G.; Cacace, F. Seismic vulnerability assessment based on typological characteristics. The first level procedure "SAVE". *Soil Dyn. Earthq. Eng.* **2015**, *69*, 262–269. [CrossRef]
5. Crowley, H.; Colombi, M.; Borzi, B.; Faravelli, M.; Onida, M.; Lopez, M.; Polli, D.; Meroni, F.; Pinho, R. A comparison of seismic risk maps for Italy. *Bullet. Earthq. Eng.* **2009**, *7*, 149–180. [CrossRef]
6. Stucchi, M.; Meletti, C.; Montaldo, V.; Crowley, H.; Calvi, G.M.; Boschi, E. Seismic hazard assessment (2003–2009) for the Italian building code. *Bull. Seism. Soc. Am.* **2011**, *102*, 2789–2792. [CrossRef]
7. Verderame, G.; Manfredi, G.; Frunzio, G. Le proprietà meccaniche dei calcestruzzi impiegati nelle strutture in cemento armato realizzate negli anni '60. In Proceedings of the X Convegno Nazionale "L'Ingegneria Sismica in Italia", Potenza-Matera, Italy, 9–13 September 2001; pp. 9–13. (In Italian)
8. Verderame, G.; Stella, A.; Cosenza, E. Le proprietà meccaniche degli acciai impiegati nelle strutture in ca realizzate negli anni'60. In Proceedings of the X Convegno Nazionale L'Ingegneria Sismica in Italia, Potenza-Matera, Italy, 9–13 September 2001. (In Italian)
9. Masi, A.; Santarsiero, G.; Chiauzzi, L. Development of a seismic risk mitigation methodology for public buildings applied to the hospitals of Basilicata region (Southern Italy). *Soil Dyn. Earthq. Eng.* **2014**, *65*, 30–42. [CrossRef]
10. Ferrini, M.; Signorini, N.; Pelliccia, P.; Pistola, F.; Prestifilippo, V.; Sabia, G. Risultati delle campagne d'indagine svolte dalla Regione Toscana per la valutazione della resistenza del calcestruzzo di edifici esistenti in cemento armato. In Proceedings of the Conference Valutazione e Riduzione della Vulnerabilità Sismica di Edifici in Cemento Armato, Rome, Italy, 29–30 May 2008. (In Italian).
11. Masi, A.; Digrisolo, A.; Santarsiero, G. Concrete strength variability in Italian RC buildings: Analysis of a large database of core tests. *Appl. Mech. Mater.* **2014**, *597*, 283–290. [CrossRef]

12. Cristofaro, M.; D'Ambrisi, A.; De Stefano, M.; Pucinotti, R.; Tanganelli, M. Studio Sulla Dispersione dei valori di resistenza a compressione del calcestruzzo di edifici esistenti. *Il Giornale Delle Prove non Distruttive Monitoraggio Diagnostica* **2012**, *2*, 32–39.
13. Shimizu, Y.; Hirosawa, M.; Zhou, J. Statistical analysis of concrete strength in existing reinforced concrete buildings in Japan. In Proceedings of the 12WCEE 2000: 12th World Conference on Earthquake Engineering, Auckland, New Zealand, 30 January–4 February 2000.
14. Szilágyi, K.; Borosnyói, A.; Zsigovics, I. Extensive statistical analysis of the variability of concrete rebound hardness based on a large database of 60 years experience. *Construct. Build. Mater.* **2014**, *53*, 333–347. [CrossRef]
15. Masi, A.; Chiauzzi, L. An experimental study on the within-member variability of in situ concrete strength in RC building structures. *Construct. Build. Mater.* **2013**, *47*, 951–961. [CrossRef]
16. Xu, T.; Li, J. Assessing the spatial variability of the concrete by the rebound hammer test and compression test of drilled cores. *Construct. Build. Mater.* **2018**, *188*, 820–832. [CrossRef]
17. Masi, A.; Chiauzzi, L.; Manfredi, V. Criteria for identifying concrete homogeneous areas for the estimation of in-situ strength in RC buildings. *Construct. Build. Mater.* **2016**, *121*, 576–587. [CrossRef]
18. Pereira, N.; Romão, X. Assessing concrete strength variability in existing structures based on the results of NDTs. *Construct. Build. Mater.* **2018**, *173*, 786–800. [CrossRef]
19. Alwash, M.; Sbartaï, Z.M.; Breysse, D. Non-destructive assessment of both mean strength and variability of concrete: A new bi-objective approach. *Construct. Build. Mater.* **2016**, *113*, 880–889. [CrossRef]
20. Presidenza del Consiglio dei Ministri. *OPCM 3274 e s.m.i.—Allegato 2 Norme Tecniche per il Progetto, la Valutazione e L'Adeguamento Sismico Degli Edifici*; Gazzetta Ufficiale: Rome, Italy, 2003. (In Italian)
21. Building Seismic Safety Council. *NEHRP Commentary on the Guidelines for the Seismic Rehabilitation of Buildings*; FEMA Publication 274; FEMA: Washington, DC, USA, 1997.
22. Collepardi, M. *Il Nuovo Calcestruzzo*, 2nd ed.; Edizioni Tintoretto: Rome, Italy, 2002. (In Italian)
23. Masi, A.; Digrisolo, A.; Santarsiero, G. Experimental evaluation of drilling damage on the strength of cores extracted from RC buildings. *World Acad. Sci. Eng. Technol. (WASET)* **2013**, *7*, 749.
24. European Committee for Standardization. *Eurocode 8: Design of structures for earthquake resistance—Part 3: Assessment and retrofitting of buildings*; EN 1998-3:2005; European Committee for Standardization (CEN): Brussels, Belgium, 2005.
25. Fiore, A.; Porco, F.; Uva, G.; Mezzina, M. On the dispersion of data collected by in situ diagnostic of the existing concrete. *Construct. Build. Mater.* **2013**, *47*, 208–217. [CrossRef]
26. Cristofaro, M.; Pucinotti, R.; Tanganelli, M.; de Stefano, M. *The dispersion of concrete compressive strength of existing buildings. Computational Methods, Seismic Protection, Hybrid Testing and Resilience in Earthquake Engineering*; Springer: Cham, Germany, 2015; pp. 275–285.
27. American Society of Civil Engineering. *Prestandard and Commentary for the Seismic Rehabilitation of Buildings*; FEMA 356; FEMA: Washington, DC, USA, 2000.
28. Masi, A.; Dolce, M.; Caterina, F. Seismic response of irregular multi-storey buildings with flexible inelastic diaphragms. *Struct. Des. Tall Build.* **1997**, *6*, 99–124. [CrossRef]
29. De Stefano, M.; Tanganelli, M.; Viti, S. Effect of the variability in plan of concrete mechanical properties on the seismic response of existing rc framed structures. *Bullet. Earthq. Eng.* **2013**, *11*, 1049–1060. [CrossRef]

© 2019 by the authors. Licensee MDPI, Basel, Switzerland. This article is an open access article distributed under the terms and conditions of the Creative Commons Attribution (CC BY) license (http://creativecommons.org/licenses/by/4.0/).

Article

Quantifying the Adhesion of Silicate Glass–Ceramic Coatings onto Alumina for Biomedical Applications

Francesco Baino [1,2]

1. Institute of Materials Physics and Engineering, Applied Science and Technology Department, Politecnico di Torino, Corso Duca degli Abruzzi 24, 10129 Torino, Italy; francesco.baino@polito.it; Tel.: +39-011-090-4668
2. Interuniversity Center for the promotion of the 3Rs Principles in Teaching and Research, Italy

Received: 11 May 2019; Accepted: 27 May 2019; Published: 30 May 2019

Abstract: Deposition of bioactive glass or ceramic coatings on the outer surface of joint prostheses is a valuable strategy to improve the osteointegration of implants and is typically produced using biocompatible but non-bioactive materials. Quantifying the coating–implant adhesion in terms of bonding strength and toughness is still a challenge to biomaterials scientists. In this work, wollastonite ($CaSiO_3$)-containing glass–ceramic coatings were manufactured on alumina tiles by sinter-crystallization of SiO_2–CaO–Na_2O–Al_2O_3 glass powder, and it was observed that the bonding strength decreased from 34 to 10 MPa as the coating thickness increased from 50 to 300 µm. From the viewpoint of bonding strength, the coatings with thickness below 250 µm were considered suitable for biomedical applications according to current international standards. A mechanical model based on quantized fracture mechanics allowed estimating the fracture toughness of the coating on the basis of the experimental data from tensile tests. The critical strain energy release rate was also found to decrease from 1.86 to 0.10 J/m^2 with the increase of coating thickness, which therefore plays a key role in determining the mechanical properties of the materials.

Keywords: biomaterials; bioceramics; coating; mechanical properties

1. Introduction

Deposition of glass or ceramic coatings on orthopedic and dental implants is useful to impart better biocompatibility and bone-bonding properties [1] to the underlying material, which typically exhibits high mechanical performance but is not inherently bioactive (e.g., titanium and its alloys [2,3]). Furthermore, the coatings have a protective function, reducing implant corrosion (especially in the case of metallic implants) and protecting living tissues against corrosion products [4].

Plasma-sprayed hydroxyapatite coatings have been used for many years on titanium implants in orthopedic and dental surgery [5]; however, the rise of some concerns about the stability of the calcium phosphate layer (interfacial delamination) has stimulated the search for new options, like bioactive glasses. These materials have the unique capability of forming a tight bond with bone and soft collagenous tissues through a series of chemical reactions conceptually similar to the mechanism of conventional glass corrosion [6]. Hence, bioactive glasses are, by nature, reactive in aqueous solution and are prone to degrade over time according to a wide range of dissolution kinetics that strongly depend on the glass composition. The risk of uncontrolled dissolution, carrying the problem of prosthesis mobility in the long term [7], is perhaps the major reason why the application of bioactive glass coatings on permanent implants is still very limited in clinics and, to date, an ideal bioactive material for coating does not exist, which motivates further research.

The literature on biomedical glass coatings is mainly focused on both the search for effective deposition methods, which evolved from conventional enameling [8] to electrophoretic deposition [9] and laser-based methods [10], and the evaluation of coating degradation, biocompatibility, and osteointegration [11]. The mechanical performance of the coating, being quite difficult to quantitatively assess, has often been neglected. The crack path resulting from interfacial indentation was studied in order to evaluate coating resistance; however, this approach does not allow the adhesion strength to be quantitatively measured, thereby making unreliable the comparison between different systems [12,13]. Fracture toughness has also been estimated by measuring the crack length in indentation tests; however, these methods may be prone to significant errors due to violations of model assumptions during practical testing situations, especially when brittle coatings are tested [14,15]. Overall, the existing literature witnesses that quantifying the coating adhesion and toughness is still a challenge but should be crucial in the development of new glass coatings.

In this work, it is studied how thickness can affect the mechanical performance of silicate glass–ceramic coatings deposited onto alumina. To the best of the author's knowledge, it is the first time that such a combination of coating material (SiO_2–CaO–Na_2O–Al_2O_3 base glass) and substrate (alumina) is quantitatively studied from this viewpoint. Specifically, the bonding strength of the coatings was experimentally assessed by performing tensile tests, and the fracture toughness was then determined by applying a properly derived model based on quantized fracture mechanics.

2. Materials and Methods

2.1. Glass Preparation

The starting silicate glass ($57SiO_2$–$34CaO$–$6Na_2O$–$3Al_2O_3$ wt.%), previously developed for biomedical applications [16], was produced by a melting-quenching route. Firstly, the reagents (analytical-grade powders of SiO_2, $CaCO_3$, Na_2CO_3, and Al_2O_3, all purchased from Sigma-Aldrich, St. Louis, MO, USA) were homogenously mixed for 30 min in an orbital shaker; then, the blend of powders was transferred to a platinum crucible, heated to 1150 °C (heating rate 10 °C/min) in an electrically heated furnace, and melted for 1 h in air. The melt was quenched into deionized water to obtain a frit that was ball-milled (Pulverisette 0, Fritsch, Idar-Oberstein, Germany) and sieved by a stainless-steel sieve (Giuliani Technology Srl, Turin, Italy) to a final particle size below 32 μm.

2.2. Coating Production

Glass coatings of different nominal thicknesses (50, 100, 150, 200, 250, and 300 μm) were manufactured on flat high-purity alumina tiles (size 10 mm × 10 mm, thickness 1 mm; produced by diamond-cutting of 200 mm × 200 mm AL603103 99% Al_2O_3 sheets, Goodfellow, Coraopolis, PA, USA) by gravity-guided deposition after suspending proper amounts of glass powder in ethanol. The glass particles were left to deposit on the substrates overnight; then, the containers hosting the materials were placed in an oven at 90 °C for 6 h to remove the excess ethanol and allow complete dying of the "greens". The samples were then thermally treated in an electrically heated furnace at 1000 °C for 3 h (heating rate 5 °C/min) in order to allow glass particle sintering and coating consolidation.

2.3. Characterizations

Differential thermal analysis (DTA) was performed on the glass powder to determine the characteristic temperatures of the material, i.e., glass transition temperature (T_g), crystallization onset (T_x), crystallization peak (T_p), and melting (T_m). Specifically, glass powder was analyzed by using a DTA 404 PC instrument (Netzsch, Selb, Germany) in the temperature range of 25 to 1400 °C with a heating rate of 10 °C/min. After being introduced into a platinum crucible provided by the instrument manufacturer, 50 mg of glass underwent the thermal cycle; an equal amount of high-purity Al_2O_3 powder was put in the reference crucible. Standard calibration procedure and baseline corrections were performed.

Sintered coatings underwent X-ray diffraction (XRD) in the 2θ-range of 10 to 60° by using a X'Pert Pro PW3040/60 diffractometer (PANalytical, Eindhoven, The Netherlands) operating at 40 kV, 30 mA with Cu Kα incident radiation (wavelength 0.15405 nm), step size 0.02°, and counting time

1 s per step. Crystalline phase assignment was performed by using X'Pert HighScore software 2.2b (PANalytical, Eindhoven, The Netherlands) equipped with the PCPDFWIN database (http://pcpdfwin.updatestar.com).

The coating cross sections and the substrate–coating interface were also inspected by scanning electron microscopy (SEM). For this purpose, the samples were embedded in epoxy resin (Epofix, Struers, Ballerup, Denmark), cut by a diamond wheel (Accutom, Struers, Ballerup, Denmark), and polished using #600 to #4000 SiC grit paper; the resulting cross sections were sputter-coated with silver and analyzed by field-emission SEM (Supra^{TM} 40, Zeiss, Oberkochen, Germany) at an accelerating voltage of 15 kV. The thickness of the coating was measured directly on the SEM images by making use of an image-analysis software provided by the manufacturer; the results were expressed as average ± standard deviation of three measurements per each sample. Compositional analysis was performed by energy dispersive spectroscopy (EDS); the probe was included in the SEM equipment.

The tensile (bonding) strength was measured by applying tensile loads to pull the coating apart, following the relevant ASTM standard [17]. Before being tested using an MTS machine (cross-head speed 1 mm/min), each sample was glued to two stainless-steel cylindrical fixtures (diameter 16 mm) by a strong bi-component adhesive (Araldite® AV 119, Huntsman, Woodlands, TX, USA), according to an experimental procedure described elsewhere [16]. The bonding strength (σ) was calculated as the maximum load (F) per unit cross-sectional area (A):

$$\sigma = \frac{F}{A} \tag{1}$$

Pull-out tests were performed on five specimens per each of the six sample batches having different nominal thickness; the results were expressed as average ± standard deviation.

3. Toughness Estimation: Derivation of the Mechanical Model

An estimation of the coating toughness was obtained from the values of tensile strength (see Equation (1)) by implementing an appropriate model, based on fracture mechanics, which was derived as follows. The theory of linear elastic fracture mechanics states that the total potential energy of a system can be expressed as:

$$\Pi = U - W \tag{2}$$

where U is the strain energy, and W is the work done by the external force F. Considering our case, U and W can be calculated as:

$$U = \frac{1}{2}F^2\left(\frac{1}{k_1} + \frac{1}{k_2}\right) \tag{3}$$

$$W = F^2\left(\frac{1}{k_1} + \frac{1}{k_2}\right) \tag{4}$$

where k_1 and k_2 are the stiffness of alumina tile and coating, respectively, before crack propagation.

Hence, after combining Equations (2)–(4), it is obtained:

$$\Pi = -\frac{1}{2}F^2\left(\frac{1}{k_1} + \frac{1}{k_2}\right) \tag{5}$$

Griffith's energy criterion states that crack propagation occurs when the variation of the total potential energy $d\Pi$, corresponding to a virtual increment of crack surface dA, is equal to the energy needed to create the new free crack surface, i.e., $d\Pi = -G_{IC}dA$, where G_{IC} is the fracture energy (per unit area created) of the material.

The recent theory of quantized fracture mechanics extends the Griffith's criterion to discrete cracks: thus, after substituting differentials with finite differences (i.e., $d \to \Delta$), the Griffith's equation can be rewritten as [18]:

$$G_{IC} = -\frac{\Delta \Pi}{\Delta A} \quad (6)$$

There are three possible failure modes occurring when the coatings undergo pull-out tests: in fact, the fracture can be (i) totally adhesive, (ii) totally cohesive, or (iii) mixed.

Considering the case (i), it is ideally assumed that the failure occurs at the coating–alumina interface; therefore, the variation of the total potential energy is:

$$\Delta \Pi = -\frac{1}{2}F^2\left[\left(\frac{1}{k_1^*} - \frac{1}{k_1}\right) + \left(\frac{1}{k_2^*} - \frac{1}{k_2}\right)\right] \quad (7)$$

where k_1^* and k_2^* are the stiffness after crack propagation.

The compliances in Equation (7) can be expressed as $\frac{1}{k_1^*} - \frac{1}{k_1} = \frac{t_1}{E_1 A^2} \cdot \frac{\Delta A}{1-\left(\frac{\Delta A}{A}\right)}$ and $\frac{1}{k_2^*} - \frac{1}{k_2} = \frac{t_2}{E_2 A^2} \cdot \frac{\Delta A}{1-\left(\frac{\Delta A}{A}\right)}$, where E_1 (=400 GPa) and E_2 (=90 GPa [19]) are the Young's moduli of alumina and coating, respectively, and t_1 (=1 mm) and t_2 the corresponding thicknesses.

Thus, Equations (7) can be rewritten as:

$$\Delta \Pi = -\frac{1}{2}F^2\left(\frac{t_1}{E_1 A^2} \cdot \frac{\Delta A}{1-\left(\frac{\Delta A}{A}\right)} + \frac{t_2}{E_2 A^2} \cdot \frac{\Delta A}{1-\left(\frac{\Delta A}{A}\right)}\right) \quad (8)$$

Combining Equations (6) and (8), the energy release rate $G_{I,12}$ can be expressed as:

$$G_{I,12} = (\sigma_{I,12})^2 \frac{E_1 t_2 + E_2 t_1}{2E_1 E_2 \left(1 - \left(\frac{\Delta A}{A}\right)\right)} \quad (9)$$

Hence, the delamination (critical) strength can be expressed as:

$$\sigma_{IC,12} = \sqrt{\frac{2E_1 E_2}{E_1 t_2 + E_2 t_1} G_{IC,12}\left(1 - \frac{\Delta A}{A}\right)} \quad (10)$$

Considering the case (ii), it is ideally assumed that the failure occurs entirely in the coating; this means that the tensile strength of the coating material ($\sigma_{IC,2}$ = 47 MPa [19]) is lower than the adhesion strength at the coating–substrate interface.

A simple visual inspection of the fracture surfaces after tensile tests revealed that there was a coexistence of the two failure modes (i) and (ii) (adhesive + cohesive, resulting in mode (iii)). Therefore, the critical stress is assumed to be predicted by a mean field approach as:

$$\sigma_{IC} = \sigma = \sigma_{IC,12}\frac{A_d}{A} + \sigma_{IC,2}\left(1 - \frac{A_d}{A}\right) \quad (11)$$

where $A_d = A - \Delta A$ is the delamination area.

After combining Equations (10) and (11), it is finally obtained:

$$G_{IC,12} = \frac{\left[\sigma - \sigma_{IC,2}\left(1 - \frac{A_d}{A}\right)\right]^2}{\frac{2E_1 E_2}{E_1 t_2 + E_2 t_1}\left(\frac{A_d}{A}\right)^3} \quad (12)$$

4. Results and Discussion

The silicate glass used in this work was initially proposed for potential usage in the field of bone regeneration and substitution [16]. This composition was then discarded with respect to such applications, as the high content of SiO_2 and the significant presence of Al_2O_3 resulted in an almost inert behavior with almost no apatite-forming capability in contact with biological fluids, thereby preventing the formation of a tight bond to bone. However, the material was found highly suitable to produce porous orbital implants [20], which must elicit minimal or no reactions in contact with ocular tissues [21], and multilayer coatings onto ceramic implants [22].

The characteristic temperatures of the glass, determined from the DTA plot (Figure 1), were $T_g = 685\ °C$ (inflection point), $T_x = 800\ °C$ (crystallization onset), $T_p = 855\ °C$ (exothermic peak), and $T_m = 1155\ °C$ (endothermic peak).

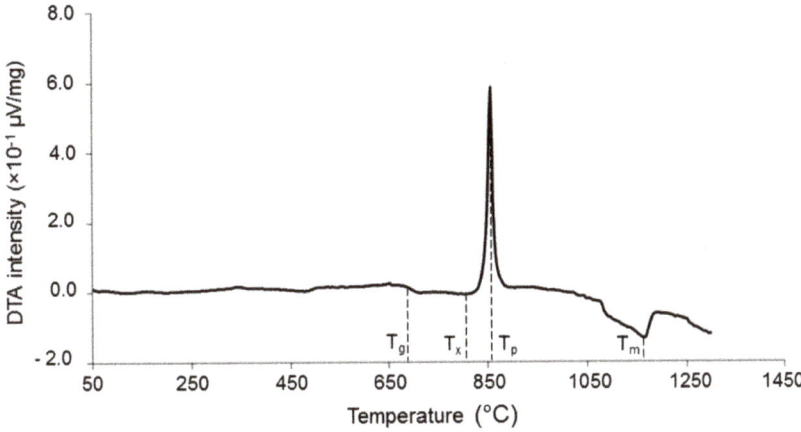

Figure 1. DTA plot of the starting glass powder used to manufacture the coatings.

The coatings underwent devitrification upon thermal treatment, as shown in Figure 2. The development of wollastonite crystals ($CaSiO_3$, PDF code no. 00-027-0088) when the material was thermally treated at 1000 °C is consistent with the results from thermal analysis, showing that crystallization started at 855 °C. The XRD results were also in agreement with those obtained on the same glass system after heat treatment above 900 °C [23]. No problems of possible cytotoxicity due to wollastonite were forecast, as this crystalline phase was clearly proved to be highly biocompatible and suitable in contact with living bone, also for filling load-bearing osseous defects [24].

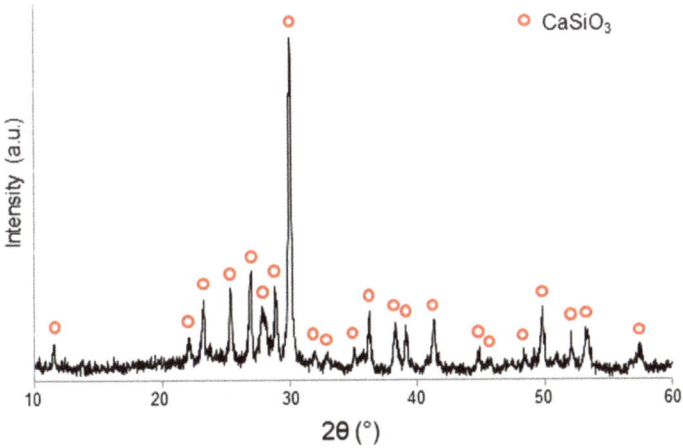

Figure 2. XRD pattern of a sintered coating (1000 °C/3 h).

Analysis of the cross section revealed that the interface between coating and alumina tile was flawless without any apparent interfacial crack or delamination in intact samples (Figure 3a). However, some isolated small pores could be observed in the coating. The quality of adhesion between coating and alumina tile after thermal treatment was very good due to an almost perfect matching between the thermal expansion coefficients of glass (8.7×10^{-6} °C^{-1} [16]) and alumina (8.5×10^{-6} °C^{-1}).

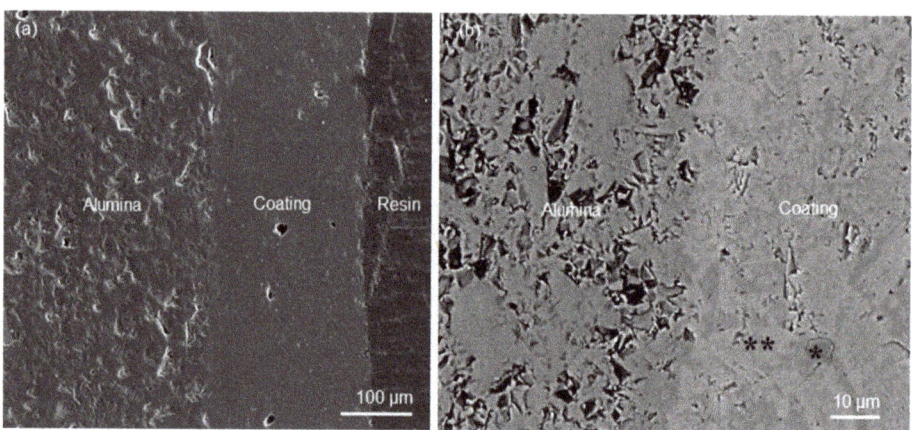

Figure 3. Analysis of the coating cross section: (**a**) SEM micrograph in "secondary" mode (400×) and (**b**) high-magnification detail (2500×) inspected in back-scattering mode showing the interface between coating and alumina as well as the coating microstructure (glass–ceramic nature); the white areas (**) correspond to the wollastonite crystals, while the dark zones (*) correspond to the residual glassy phase.

The thicknesses of the coatings, reported in Table 1, were in good agreement with the nominal values. The slight decrease compared to the nominal thickness can be attributed to the glass powder densification that occurred during sintering.

Table 1. Mechanical properties of the coatings.

Nominal/Measured Thickness of the Coating (µm)	A_d (mm^2)	σ (MPa)	$G_{IC,12}$ (J/m^2)
50/48 ± 5	92.0 ± 4.3	34.4 ± 3.1	1.86 ± 0.35
100/94 ± 6	85.0 ± 3.7	28.0 ± 1.6	1.29 ± 0.19
150/142 ± 8	86.3 ± 6.0	21.3 ± 2.6	0.72 ± 0.29
200/194 ± 10	72.0 ± 2.5	17.4 ± 1.1	0.12 ± 0.063
250/240 ± 8	71.2 ± 8.4	14.2 ± 3.2	0.11 ± 0.013
300/296 ± 8	71.4 ± 3.6	10.0 ± 1.2	0.10 ± 0.076

The SEM micrograph shown in Figure 3b, acquired in back-scattering mode, clearly shows the glass–ceramic nature of the coating, in which wollastonite crystals and residual glassy phase coexisted. This was further confirmed by the compositional analyses (EDS): in fact, the CaSiO$_3$ crystals ("white" areas with atomic composition 43.2% Si, 56.8% Ca), exhibiting a typical acicular shape, were embedded in a grey matrix in which high amounts of Al and Na could be found, too, along with low Ca content (atomic composition 22.6% Na, 13.1% Al, 58.9% Si, 5.4% Ca), corresponding to the residual glass. These results are in accordance with the XRD pattern (Figure 2), where a bump in the 2θ-range of 20 to 35° can be observed, which is the typical proof of the presence of an amorphous phase. The back-scattering SEM image was analyzed by the free software ImageJ version 2016 (https://imagej.nih.gov/ij/index.html) to obtain a rough quantification of crystalline and glassy phase proportions in the glass–ceramic material: the white regions corresponding to CaSiO$_3$ crystals were estimated to be about 70% of the total cross-sectional area of the coating.

The results from the tensile tests showed that the bonding strength decreased as the coating thickness increased (Table 1). Data interpolation by the least-squares method suggested a linear relationship between bonding strength and thickness of the coating (Figure 4); the high value of the coefficient R^2 also confirmed the good fitting of the interpolating function with the experimental data.

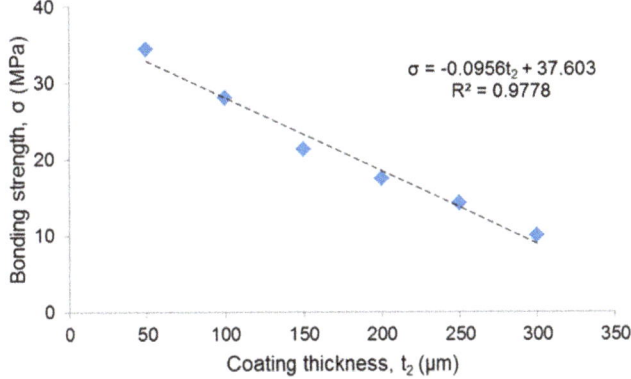

Figure 4. Relationship between bonding strength and coating thickness (blue points: experimental values, dashed line: fitting curve).

No specific international standard exists about the minimum bonding strength recommended for biomedical glass coatings; however, a tensile strength of 15 MPa is prescribed by ISO 13779 for hydroxyapatite coatings on surgical implants [25]. This requirement was fulfilled by the coatings produced in this work with thickness below 250 μm (Table 1).

Analysis of the fracture surfaces (Figure 5) confirmed a mixed failure mode (adhesive + cohesive failure), consistently with the assumption behind Equation (11). The critical strain energy release rate assessed by applying Equation (12) decreased as the coating thickness increased (Table 1). Since the $G_{IC,12}$ value can be interpreted as an estimation of fracture toughness [18], this means that the thicker coatings were less tough than the thin ones. This trend is consistent with previous observations reported in the literature about glass and ceramic coatings deposited onto metallic implants, albeit a clear quantification of the trend has been seldom reported. Furthermore, to the best of the author's knowledge, no study has been specifically conducted to determine the toughness–thickness relationship in glass coatings deposited on bioceramic substrates, which was attempted for the first time in the present work.

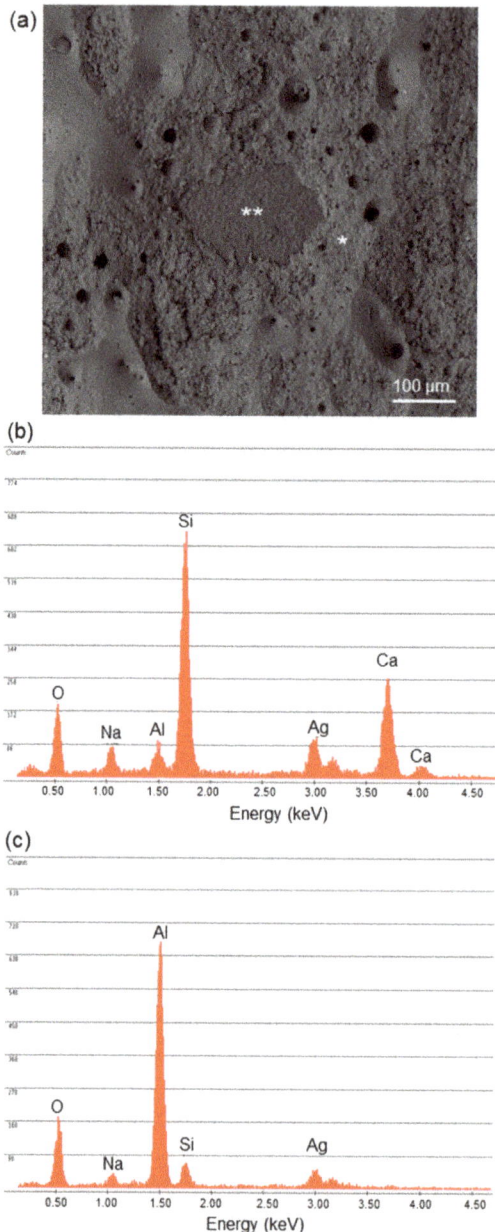

Figure 5. Analysis of fracture surface: (**a**) SEM micrograph showing the morphology of the coating after tensile test and (**b**,**c**) compositional assessment to identify the areas of adhesive and cohesive fracture. The region marked with (*) in (**a**) corresponds to the fracture area inside the coating (the corresponding EDS pattern in (**b**) reveals the presence of all the typical elements of the glass composition, i.e., Si, Ca, Na, and Al), while the region marked with (**) corresponds to interfacial delamination (a high peak for Al is visible in the corresponding EDS pattern in (**c**), with just small traces of Si and Na).

Gomez-Vega et al. [12,26] claimed that thinner silicate bioactive glass and composite coatings on titanium implants were significantly less prone to interfacial cracking but did not provide any quantitative assessment of bonding strength and fracture toughness. Zhao et al. [10] produced bioactive SiO_2–CaO–Na_2O–P_2O_5–MgO glass coatings by pulsed laser deposition on Ti6Al4V alloy and observed that thick coatings showed poorer adhesion to the substrate. Matinmanesh et al. [27] reported a similar trend for SiO_2–CaO–Na_2O–P_2O_5–ZnO coatings on Ti6Al4V alloy, too, and demonstrated that increments in the coating thickness led to higher residual stresses that tended to increase the "crack driving force". This interesting explanation can be extended to the present work. The residual stresses in the coating contributed to the release of the stored strain energy during the tensile test and, therefore, less energy from the applied loads was needed to reach the critical value required for crack propagation. In other words, lower tensile loads were required to overcome the resistance of thicker coatings against cracking because of higher residual stresses.

Wollastonite-containing glass–ceramic coatings (thickness 140 µm) deposited on alumina by airbrushing in a previous work exhibited a fracture toughness of 0.8 J/m^2 [23], which is very close to the value reported in Table 1 for the 150 µm-thick samples.

Data interpolation by the least-squares method suggested a quadratic (polynomial) relationship between fracture toughness and thickness of the coating (Figure 6), as confirmed by the high value of the coefficient R^2. A significant drop of $G_{IC,12}$ values accompanied by a decrease of A_d (Table 1) could be observed if the coating thickness exceeded 150 µm, which could be considered as a threshold value to take into account at the design stage of the coating.

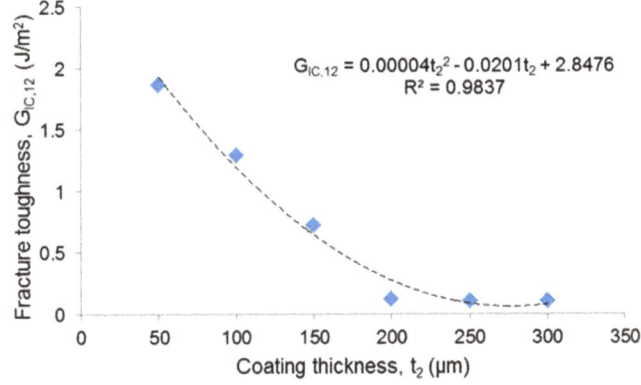

Figure 6. Relationship between fracture toughness and coating thickness (blue points: experimental values, dashed line: fitting curve).

5. Conclusions

The bonding strength and critical strain energy release rate of wollastonite-containing glass–ceramic coatings were found to decrease as the coating thickness increased, i.e., thinner coatings proved to have higher resistance against fracture than thicker ones. These relationships were quantified by combining experimental data from tensile tests on the coatings and a model based on quantized fracture mechanics. The obtained functions are very valuable to link the coating thickness, which can be tailored at the design and manufacturing stages, with those key mechanical characteristics. Specifically, it could be predicted what is the maximum thickness allowed so that the coating can exhibit the mechanical characteristics required for a given application. Therefore, the coating thickness should be carefully taken into account in order to optimize the material performance and design. The results also suggest that, ideally, the coating should be as thin as possible for increasing the mechanical performance; thus, further improvements are forecast if the coating thickness decreases below the minimum value (50 µm) considered in this study. The same methodological approach followed to estimate toughness could also be applied to different glass or ceramic coatings based on other compositions of biomedical interest.

Funding: This research received no external funding.

Conflicts of Interest: The authors declare no conflict of interest relevant to this article.

References

1. McEntire, B.J.; Bal, B.S.; Rahaman, M.N.; Chevalier, J.; Pezzotti, G. Ceramics and ceramic coatings in orthopaedics. *J. Eur. Ceram. Soc.* **2015**, *35*, 4327–4369. [CrossRef]
2. Chen, X.; Li, H.S.; Yin, Y.; Feng, Y.; Tan, X.W. Macrophage proinflammatory response to the Ti alloy equipment in dental implantation. *Genet. Mol. Res.* **2015**, *14*, 9155–9162. [CrossRef] [PubMed]
3. Gibon, E.; Amanatullah, D.F.; Loi, F.; Pajarinen, J.; Nabeshima, A.; Yao, Z.; Hamadouche, M.; Goodman, S.B. The biological response to orthopaedic implants for joint replacement: Part I: Metals. *J. Biomed. Mater. Res. B* **2017**, *105*, 2162–2173. [CrossRef]
4. Sola, A.; Bellucci, D.; Cannillo, V.; Cattini, A. Bioactive glass coatings: A review. *Surf. Eng.* **2011**, *27*, 560–572. [CrossRef]
5. Sun, L.; Berndt, C.C.; Gross, K.A.; Kucuk, A. Material fundamentals and clinical performance of plasma-sprayed hydroxyapatite coatings: A review. *J. Biomed. Mater. Res. (Appl. Biomater.)* **2001**, *58*, 570–592. [CrossRef] [PubMed]
6. Fernandes, H.R.; Gaddam, A.; Rebelo, A.; Brazete, D.; Stan, G.E.; Ferreira, J.M.F. Bioactive glasses and glass-ceramics for healthcare applications in bone regeneration and tissue engineering. *Materials* **2018**, *11*, 2530. [CrossRef]
7. Alonso-Barrio, J.A.; Sanchez-Herraez, S.; Fernandez-Hernandez, O.; Betegon-Nicolas, J.; Gonzalez-Fernandez, J.J.; Lopez-Sastre, A. Bioglass-coated femoral stem. *J. Bone Jt. Surg.* **2004**, *86B* (Suppl. II), 138.
8. Peddi, L.; Brow, R.K.; Brown, R.F. Bioactive borate glass coatings for titanium alloys. *J. Mater. Sci. Mater. Med.* **2008**, *19*, 3145–3152. [CrossRef]
9. Boccaccini, A.R.; Keim, S.; Ma, R.; Li, Y.; Zhitomirsky, I. Electrophoretic deposition of biomaterials. *J. R. Soc. Interface* **2010**, *7*, S581–S613. [CrossRef]
10. Zhao, Y.; Song, M.; Liu, J. Characteristics of bioactive glass coatings obtained by pulsed laser deposition. *Surf. Interface Anal.* **2008**, *40*, 1463–1468. [CrossRef]
11. Baino, F.; Verné, E. Glass-based coatings on biomedical implants: A state-of-the-art review. *Biomed. Glasses* **2017**, *3*, 1–17. [CrossRef]
12. Gomez-Vega, J.M.; Saiz, E.; Tomsia, A.P.; Marshall, G.W.; Marshall, S.J. Bioactive glass coatings with hydroxyapatite and bioglass particles on Ti-based implants. 1. Processing. *Biomaterials* **2000**, *21*, 105–111. [CrossRef]
13. Lopez-Esteban, S.; Saiz, E.; Fujino, S.; Oku, T.; Suganuma, K.; Tomsia, A.P. Bioactive glass coatings for orthopedic metallic implants. *J. Eur. Ceram. Soc.* **2003**, *23*, 2921–2930. [CrossRef]
14. Kruzic, J.J.; Kim, D.K.; Koester, K.J.; Ritchie, R.O. Indentation techniques for evaluating the fracture toughness of biomaterials and hard tissues. *J. Mech. Behav. Biomed. Mater.* **2009**, *2*, 384–395. [CrossRef]
15. Hsiung, C.H.H.; Pyzik, A.J.; Gulsoy, E.B.; De Carlo, F.; Xiao, X.; Faber, K.T. Impact of doping on the mechanical properties of acicular mullite. *J. Eur. Ceram. Soc.* **2013**, *33*, 1955–1965. [CrossRef]
16. Vitale-Brovarone, C.; Baino, F.; Tallia, F.; Gervasio, C.; Verné, E. Bioactive glass-derived trabecular coating: A smart solution for enhancing osteointegration of prosthetic elements. *J. Mater. Sci. Mater. Med.* **2012**, *23*, 2369–2380. [CrossRef]
17. ASTM C633. Standard Test Method for Adhesion or Cohesion Strength of Thermal Spray Coatings. 2017. Available online: https://www.astm.org/Standards/C633.htm (accessed on 29 May 2019).
18. Pugno, N.; Ruoff, R. Quantized fracture mechanics. *Philos. Mag.* **2004**, *84*, 2829–2845. [CrossRef]
19. Chen, Q.; Baino, F.; Pugno, N.M.; Vitale-Brovarone, C. Bonding strength of glass-ceramic trabecular-like coatings to ceramic substrates for prosthetic applications. *Mater. Sci. Eng. C* **2013**, *33*, 1530–1538. [CrossRef]
20. Baino, F. Porous glass-ceramic orbital implants: A feasibility study. *Mater. Lett.* **2018**, *212*, 12–15. [CrossRef]

21. Baino, F. How can bioactive glasses be useful in ocular surgery? *J. Biomed. Mater. Res. A* **2015**, *103*, 1259–1275. [CrossRef]
22. Baino, F.; Vitale-Brovarone, C. Trabecular coating on curved alumina substrates using a novel bioactive and strong glass-ceramic. *Biomed. Glasses* **2015**, *1*, 31–40. [CrossRef]
23. Baino, F.; Vitale-Brovarone, C. Wollastonite-containing bioceramic coatings on alumina substrates: Design considerations and mechanical modelling. *Ceram. Int.* **2015**, *41*, 11464–11470. [CrossRef]
24. Kokubo, T.; Ito, S.; Sakka, S.; Yamamuro, T. Formation of a high strength bioactive glass-ceramic in the system MgO–CaO–SiO2–P_2O_5. *J. Mater. Sci.* **1986**, *21*, 536–540. [CrossRef]
25. ISO 13779-4. Implants for Surgery: Hydroxyapatite, Part 4: Determination of Coating Adhesion Strength. 2018. Available online: https://www.iso.org/standard/64619.html (accessed on 29 May 2019).
26. Gomez-Vega, J.M.; Saiz, E.; Tomsia, A.P. Glass-based coatings for titanium implant alloys. *J. Biomed. Mater. Res.* **1999**, *46*, 549–559. [CrossRef]
27. Matinmanesh, A.; Rodriguez, O.; Towler, M.R.; Zalzal, P.; Schemitsch, E.H.; Papini, M. Quantitative evaluation of the adhesion of bioactive glasses onto Ti6Al4V substrates. *Mater. Des.* **2016**, *97*, 213–221. [CrossRef]

© 2019 by the author. Licensee MDPI, Basel, Switzerland. This article is an open access article distributed under the terms and conditions of the Creative Commons Attribution (CC BY) license (http://creativecommons.org/licenses/by/4.0/).

Article

In-Situ Damage Evaluation of Pure Ice under High Rate Compressive Loading

Matti Isakov, Janin Lange, Sebastian Kilchert and Michael May *

Fraunhofer Institute for High-Speed Dynamics, Ernst-Mach-Institut, Ernst-Zermelo-Straße 4, 79104 Freiburg, Germany; matti.isakov@tuni.fi (M.I.); janin.lange@gmx.de (J.L.); sebastian.kilchert@emi.fraunhofer.de (S.K.)
* Correspondence: michael.may@emi.fraunhofer.de

Received: 19 March 2019; Accepted: 12 April 2019; Published: 15 April 2019

Abstract: The initiation and propagation of damage in pure ice specimens under high rate compressive loading at the strain rate range of $100\ \mathrm{s}^{-1}$ to $600\ \mathrm{s}^{-1}$ was studied by means of Split Hopkinson Pressure Bar measurements with incorporated high-speed videography. The results indicate that local cracks in specimens can form and propagate before the macroscopic stress maximum is reached. The estimated crack velocity was in the range of 500 m/s to 1300 m/s, i.e., lower than, but in similar order of magnitude as the elastic wave speed within ice. This gives reason to suspect that already at this strain rate the specimen is not deforming under perfect force equilibrium when the first cracks initiate and propagate. In addition, in contrast to quasi-static experiments, in the high rate experiments the specimens showed notable residual load carrying capacity after the maximum stress. This was related to dynamic effects in fractured ice particles, which allowed the specimen to carry compressive load even in a highly damaged state.

Keywords: ice; high rate loading; compressive loading; Split Hopkinson bar; in-situ fractography

1. Introduction

The impact of solid ice, such as hail, on high velocity load carrying structures, such as the leading surfaces of airplanes (wind shields or wing leading edges) may pose a serious threat to safety if the impacting particles are large enough (typically diameter 16 mm to 50 mm) and the relative velocity between them and the moving structure is high enough (>100 m/s) that impact forces are generated which are high enough to cause plastic deformation or damage to the structure [1]. Considerable efforts have been in the past taken to quantitatively predict these effects based on numerical simulations (see for example references [2,3]). One of the main questions related to this kind of work is whether ice exhibits strain rate sensitivity at high rates of loading. There are published reports [4–8] on the measurement of the high strain rate properties of ice using the Split Hopkinson Bar method, which is a well-established technique for the measurement of material responses in the strain rate region of $100\ \mathrm{s}^{-1}$ to $10{,}000\ \mathrm{s}^{-1}$ [9]. These previous works have concentrated on measuring the strain rate sensitivity of the maximum strength of ice and the analysis has been mainly carried out based on the measured macroscopic stress-strain curves.

A question which arises is how the measured high strain rate stress-strain curves relate to the damage process within the material. Recent work by Lian et al. [10] carried out with a servo-hydraulic testing machine at strain rates of $2\ \mathrm{s}^{-1}$ to $40\ \mathrm{s}^{-1}$ indicate that specimen cracking starts already before the maximum stress is reached. Ice is known to be a brittle and stochastic material, exhibiting a large scatter in strength even at quasi-static loading rates [11]. It is thus reasonable to assume that the initiation of cracking in the specimen is a stochastic process both in terms of the required stress and the location of the first crack (s). Furthermore, cracks in ice can travel at velocities in the order of 1000 m/s [10], i.e., at velocities which are relevant in the time-scale of the measurements carried out at high rates of loading. Therefore, it seems that for the correct interpretation of the high strain rate stress-strain curves, the evolving damage within the ice specimens should be followed in-situ during the experiments.

Due to the stochastic nature of ice which associated with the growth of ice crystals, this paper does not claim to provide discrete numbers for the strength of ice. Instead, this paper presents a novel consistent testing methodology for assessing rate effects related to fracture phenomena in ice using a Split Hopkinson Pressure Bar setup instrumented with in-situ high speed imaging. The obtained high resolution images clearly show that localized damage may take place in the specimen shortly after the start of loading and that extensive damage already rapidly propagates in the material before the maximum macroscopic stress is measured. The paper concludes with a discussion of the possible ramifications of these phenomena.

2. Experimental Methods

2.1. Preparation of Specimens

The ice specimens were prepared from demineralized water with the following procedure: First blocks with approximate dimensions of 400 mm × 300 mm × 50 mm were produced by letting the water freeze at −10 °C for 24 h to 48 h. This combination of a relatively high temperature and long freezing time was selected in order to facilitate the escape of any trapped air during the freezing process. The freezing was carried out in an isolating container, which effectively allowed heat flow in only one direction, i.e., the thickness direction of the block. This orientation was maintained throughout the following preparation steps so that in the final specimens the loading axis and the direction of the heat flow during freezing were coincident. After the desired block thickness was achieved, the block was removed from the container and stored at −10 °C for a maximum of one week. The grain structure of the ice was not assessed. However, the specimens are thought to be polycrystalline as this is the more natural way to grow than single crystal ice.

In the next step the blocks were mechanically cut to rectangular pieces of approximately 100 mm × 100 mm × 50 mm. These pieces were then carefully shaped to cylinders by using aluminum forms (temperature between +5 °C and +15 °C). After this step the specimens were still larger than the target dimensions, i.e., ~Ø40 mm × 35 mm. Then the specimens were cooled again for 10 min in order to prevent excessive heating of the specimens. After this the specimens were shaped to final dimensions of Ø22 mm × 20 mm by using the aluminum forms shown by Figure 1a,b. The parallelism of the specimen loading surfaces was controlled with a purpose made jig shown in Figure 1c. The final specimens were stored for a maximum of 24 h at the respective test temperature of −20 °C before testing (Figure 1d). In a part of the test series the pre-shaped cylinders were stored for 12 h at −10 °C before final shaping and subsequent testing. This extra step was not, however, observed to affect the mechanical response of the specimens.

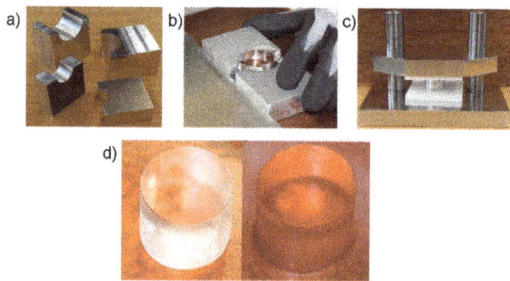

Figure 1. (**a**,**b**) aluminum forms used in the forming of the specimen; (**c**) the jig used to ensure the parallelism of the specimen loading surfaces (the aluminum cylinder at the center illustrates the specimen); (**d**) examples of specimens ready for testing.

Prior to testing, each specimen was measured with a cooled caliber and inspected for any visible damage or trapped air.

2.2. High Strain Rate Testing with the Split Hopkinson Pressure Bar

The Split Hopkinson Pressure Bar setup (Freiburg, Germany) used in the tests is schematically illustrated by Figure 2. The setup consists of an input bar (diameter 22 mm, length 2200 mm), output bar (diameter 22 mm, length 600 mm), and a striker (diameter 22 mm, length 310 mm, partially enclosed in a polymer sabot). All bars are made of high strength aluminum alloy. The input and output bars are laterally supported by stanchions, which allow for accurate alignment of the bars. The contacting parts of the stanchions are made of Teflon in order to ensure low-friction contact with the bars. Compressed air is used to accelerate the striker to the desired impact speed. In the current test series the striker impact speed was in the order of 10 m/s. A thick piece of paper was used as a pulse shaper between the striker and the input bar in order to reduce high frequency oscillations in the incident wave and to increase its rise time.

The instrumentation used in the tests involved strain gauges attached on two locations in the input bar in order to accurately characterize the wave motion during the test. On the output bar both traditional resistive strain gauges as well as semiconductor strain gauges (for improved measurement sensitivity) were attached on one location on the bar. The strain gauge signals were measured at 10 MHz with a digital oscilloscope. It should be noted that no data filtering was applied on the measurement signals. In addition to the strain gauge signals, specimen deformation and damage was monitored using a high speed camera (Phantom v1610, Wayne, NJ, USA) recording at a frame rate of 250 kHz, allowing—for the first time—detailed in-situ observation of crack initiation and propagation in dynamically loaded ice specimens. Accurate synchronization between the oscilloscope and the camera was obtained using a trigger/feedback signal-loop. Preliminary tests carried out on an aluminum specimen with dimensions similar to the specimens (length 20 mm, diameter 22 mm) were used to verify the synchronization between the two devices (Figure 3). In order to assess the validity of the wave analysis, a black-and-white speckle pattern was applied to the bar ends and the specimen. The commercial digital image correlation software package GOM Correlate (v17, Braunschweig, Germany) was then used to track the movement of the ends of the input bar and the output bar. A subset size of 0.5 mm × 0.5 mm was used for this analysis. It can be seen that a good correlation is achieved between DIC measurements and wave analyses.

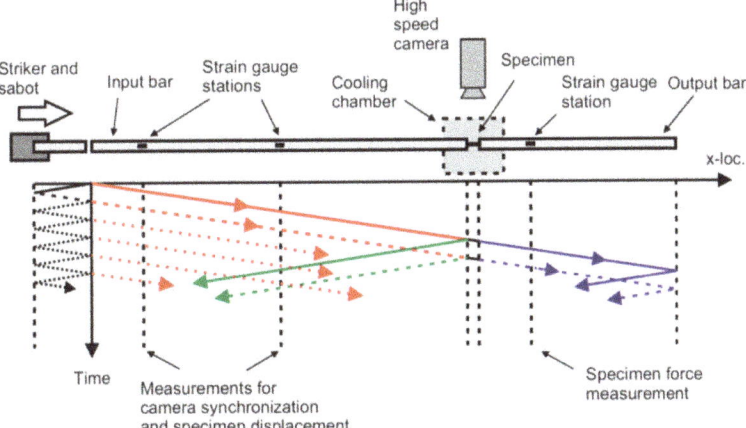

Figure 2. Schematic illustration of the components of the SHPB setup and the relevant wave motion within the bars during the test. The left hand strain gauge on the input bar is referred to as strain gauge 1, the right hand strain gauge on the input bar is referred to as strain gauge 2.

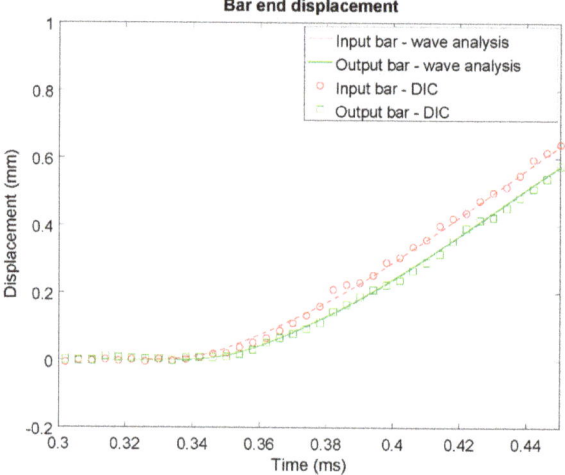

Figure 3. Illustration of the verification of the strain-gauge signal/high speed camera synchronization: bar end displacements obtained from the stress wave analysis and from the high speed footage using digital image correlation. In place of the ice specimens an aluminum cylinder with similar dimensions was used.

Specimen temperature was controlled by means of a temperature chamber made of transparent Plexiglas, which enclosed the specimen and part of the bars, as shown by Figure 4. A flow of cryogenic nitrogen gas was used to control the chamber temperature. Specimen temperature was verified immediately prior to the impact loading by means of a thermocouple placed in contact with the specimen. In order to ensure low friction contact between the bars and the specimen, Teflon sheets (thickness 0.25 mm) were placed on the contact interfaces. This introduced a challenge in holding the specimen in place before the loading. In this project a specimen support made of Teflon (visible in Figure 4 at the center of the image) was used. The lateral confinement provided by the support column is very limited. The supports are essentially with short line or almost point-contact with the specimen, i.e., there is no cylindrical confinement effect on the specimen which would cause hydrostatic stresses within the specimen. In addition, the relatively low lateral expansion of the test material during compression loading (strain to failure less than 2%) indicates that the effect of lateral confinement is low. This assumption is further justified by the high speed footage presented in Section 3 and Appendix A, which indicates that initial cracking did not take place near the support in any of the tested specimens.

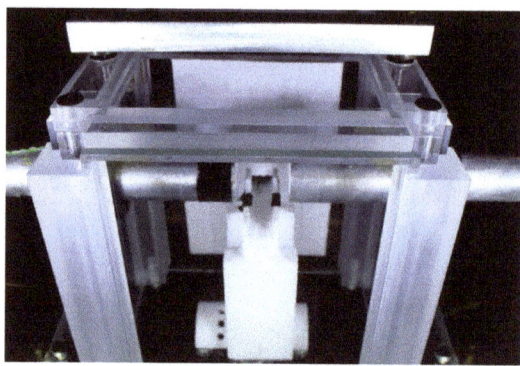

Figure 4. SHPB temperature control chamber made of transparent Plexiglas.

2.3. Data analysis of Dynamic Tests on Ice

In general, the analysis of the SHPB experiments followed the typical practices used in the field [9]. That is, the experiments involved measuring the incident wave generated by the striker into the input bar, the reflected wave originating from the input bar/specimen-interface and traveling in the input bar back towards the striker, as well as the transmitted wave originating from the specimen/output bar-interface and traveling in the output bar in the original direction. The recorded wave data was then analyzed to obtain the force-time and displacement-time response of the specimen based on the theory of uniaxial elastic stress waves in slender bars [9]. In the following section, the key points necessary for the interpretation of the results are presented. As illustrated by Figure 2, the sabot, which partially encloses the striker and remains in contact with the striker throughout the loading, introduces nonperfect release wave generation within the striker. This results in a series of oscillations following the main incident wave, as shown by Figure 5a. Unless corrected for, these trailing incident oscillations partially overlap the reflected wave at the strain gauge station. In order to establish the accurate measurement of the reflected wave, the two strain gauge stations on the input bar were used to separate the trailing oscillations of the incident wave from the reflected wave by means of deconvolution (Equation (1)):

$$\varepsilon_{st2deconv}(t) = \varepsilon_{st2}(t) - \varepsilon_{st1}(t - \Delta t) \quad (1)$$

In Equation (1) $\varepsilon_{st1}(t)$ and $\varepsilon_{st2}(t)$ refer to the strain signals measured by the first and second strain gauge station on the input bar for time t, respectively. The deconvoluted signal at the second station is denoted by $\varepsilon_{st2deconv}(t)$, which also contains the reflected wave signal from the bar-to-specimen interface. The travel time of the longitudinal elastic wave between the two stations is denoted by Δt.

Figure 5b shows an example of the incident, reflected and transmitted wave obtained after carrying out time-shifting to specimen interface location and deconvolution of the reflected wave. Careful analysis of preliminary tests carried out both in the "free end" condition (no contact at the input bar) and in the "bars together condition" (input bar in contact with the output bar without specimen) indicated that the measurements and the deconvolution process carried out for the reflected wave were not accurate enough for the reliable determination of the force acting at the input bar/specimen interface due to the low force carried by the specimen (recall that the force on the input bar is determined as the sum of the incident and reflected waves [9], which makes it highly susceptible to any uncertainty in the measurement especially in the case when the force carried by the specimen is low). Therefore, the longitudinal force acting (F) on the specimen was determined only on the specimen/output bar interface (Equation (2)):

$$F(t) = E_{bar} A_{bar} \varepsilon_{trans}(t) \quad (2)$$

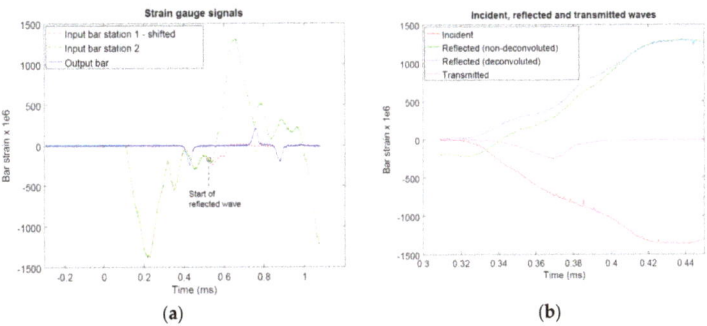

Figure 5. Example of the strain gauge signals recorded during the SHPB tests: (a) signals from the strain gauge stations and (b) incident, reflected (with and without deconvolution) and transmitted waves after time-shifting to specimen interface location (zoom-in on the early portion of the waves).

In Equation (2) E_{bar} and A_{bar} denote the Young's modulus and cross-sectional area of the bar, respectively. The transmitted wave strain amplitude (time-shifted to specimen/output bar interface) is denoted by ε_{trans}. The accuracy of the measurements and the deconvolution was determined to be sufficient for the calculation of the bar end velocities (v_{input} and v_{output}) (Equations (3) and (4)):

$$v_{input}(t) = -c_{bar}\left(\varepsilon_{inci}(t) - \varepsilon_{ref}(t)\right) \qquad (3)$$

$$v_{output}(t) = -c_{bar}\varepsilon_{trans}(t) \qquad (4)$$

In Equations (3) and (4) $\varepsilon_{inci}(t)$, $\varepsilon_{ref}(t)$, and $\varepsilon_{trans}(t)$ denote the incident, reflected (deconvoluted), and transmitted waves moved to the bar/specimen interfaces by time-shifting, respectively. The longitudinal elastic wave speed in the bar is denoted by c_{bar}. The obtained bar end velocities were integrated over time to obtain displacements and after subtraction, the specimen elongation. The wave analysis was compared with the DIC measurements carried out in a preliminary test with aluminum specimen. As shown in Figure 3, good correlation was obtained between the two measurements.

The high speed video footage of the tests was processed by using the openly available Octave software package (v4.2.1, available at https://www.gnu.org/software/octave/) in the following manner: For each test, the image frame corresponding to the specimen immediately prior to loading was selected as reference. Then each subsequent frame was processed by subtracting the corresponding pixel gray value of the reference frame from the current frame. Finally, the contrast of the processed frame was digitally enhanced by stretching the gray values to correspond to the whole available gray value range. To summarize the data analysis, the measurement data is presented in Section 3 in the following manner: (1) Macroscopic engineering stress-strain response of the specimen calculated based on the force-elongation obtained from the wave analysis and (2) comparison of the stress (force)-time signal obtained from the wave analysis with the processed high speed footage of the specimen cracking pattern. It should be noted that since the effect of the compliance of the Teflon sheets between the specimens and the bars was not corrected for, the stiffness of the test material is underestimated by the results presented here. The time and displacement axis in the presented data refer to the actual specimen loading event as obtained from above described wave analysis and synchronization between the digital oscilloscope and the high speed camera. No further adjustments of the time or strain axis were carried out in the analysis of the experimental data, unless otherwise indicated. The target strain rate for the test series was between 200 s^{-1} and 400 s^{-1}. As is seen later in the results, the instantaneous macroscopic strain rate of the specimen varies notably during the tests. This is caused by the fact that the specimen deformation and failure already took place during the phase, at which time the incident wave was still increasing in amplitude.

2.4. Quasi-Static Tests

In addition to the high rate tests, quasi-static tests were carried out on similar specimens using a ZwickRoell Z250 (Ulm, Germany) electro-mechanical materials testing machine with an incorporated temperature control chamber. Tests were carried out at a constant displacement rate of 0.05 mm/s, resulting in a nominal strain rate of 0.0025 s^{-1}. Similarly to the high rate tests, Teflon sheets (thickness 0.25 mm) were placed on the specimen/anvil-contact surfaces. The measurement data included machine load cell and displacement sensor readings collected at 100 Hz as well as video footage at 10 Hz taken with a digital video camera incorporated in the system. Specimen strain was calculated based on the displacement sensor reading. Since the effect of the compliance of the loading frame or the Teflon sheets was not corrected for, the stiffness of the test material is underestimated by the results presented here.

3. Results

A total of five valid ice tests were performed under quasi-static loading conditions and a total of six valid ice tests were performed under high-rate loading conditions. Figure 6 presents the macroscopic specimen response measured in the high rate tests at −20 °C. As is evident in Figure 6a,c, there is a large scatter in the specimen strength, which reflects the brittle and stochastic nature of the material. As noted above, the instantaneous specimen strain rate varied notably during the tests, as seen in Figure 6b. The strain rate corresponding to the maximum strength was between 200 s^{-1} and 400 s^{-1} in this test series. Despite the scatter in the strength of the specimens, all tests indicate a similar material response: First an almost linear increase of stress with respect to strain until maximum stress and then a gradual decrease in stress with secondary peaks observed in some of the tests. In some tests there is a clear initial zero-stress level in the unshifted stress-strain curve (Figure 6c), which is probably related to a small initial gap between the specimen and the bars (based on the wave analysis, this gap was estimated to be 0.15 mm or less). When the curves are systematically shifted along the strain axis to a common starting point (Figure 6d)), all tests indicate an almost linear initial stress-strain response, though with varying slopes. For quasi-static loading, the measured engineering mean peak stress was 15.6 MPa (COV 31.6%), for high-rate loading, the measured engineering mean peak stress was 12.4 MPa (COV 34.1%). Furthermore, in all high rate tests notable residual load carrying capacity was observed after the maximum stress.

Figure 7 presents the processed high speed footage for high rate tests m20_01 and m20_05 (please see supplementary material Videos S1 and S2 for the respective videos and Figures A1–A4 for respective data for the other high rate tests) alongside the synchronized specimen stress signal measured by the output bar strain gauge station. Several points are evident: The damage process does not seem to initiate in any preferred location in the specimens and visible damage already appears before the maximum stress is reached. Furthermore, the maximum stress corresponds relatively well with the time at which damage is seen throughout the specimen length, except for tests m20_04 and m20_05, in which the stress starts to decrease sooner than visible damage is seen throughout the specimen length. In the tests with the highest measured specimen strength, tests m20_01 and m20_02, the initial damage appears to take place as longitudinal cracks along the specimen, especially in test m20_01. In contrast, in tests with lower specimen strength, the damage is more diffuse. This is especially pronounced in the test with the lowest specimen strength, test m20_05, in which a diffuse and wide damage front develops at the input bar/specimen interface and propagates through the specimen volume. In this test the initial zero-stress level in the stress-strain curve (Figure 6c) was the longest, implying that the initial contact between the specimen and the bars was not perfect. This might have contributed to the damage initiating at the input bar side of the specimen. As is evident in the high speed footage, in all high rate tests the damage eventually fills the whole visible specimen volume. Comparison with stress-time data shows that under high rate loading the specimens carry load even in a highly damaged state.

Figure 6e,f present the results of the quasi-static tests carried out at −20 °C. In these tests brittle behavior after the stress maximum was observed, except for the test QS_m20_07. In tests QS_m20_01 and QS_m20_03 the stress-strain curve was smooth until maximum stress, at which point the specimen failed practically instantaneously (when compared to the time scale of the quasi-static test). The brittle failure was also clearly seen in the recorded video footage, as shown in Figure 8 and in supplementary material S3 for test QS_m20_01. Careful inspection of the quasi-static data (Figure 6f) reveals that in some of these tests (QS_m20_04, QS_m20_06, QS_m20_07) intermittent drops in specimen stress took place before the maximum stress was reached. The occurrence of the load drops could be well related to partial cracking taking place in the specimen, as shown by Figure 9 for test QS_m20_06. The full video is provided as supplementary material S4. In test QS_m20_07, in which the slope of the stress-strain curve decreases notably after the first drop in stress (Figure 6f), continuous formation of longitudinal cracks was observed before and after the point of maximum stress until the final failure took place, as shown in Figure 10.

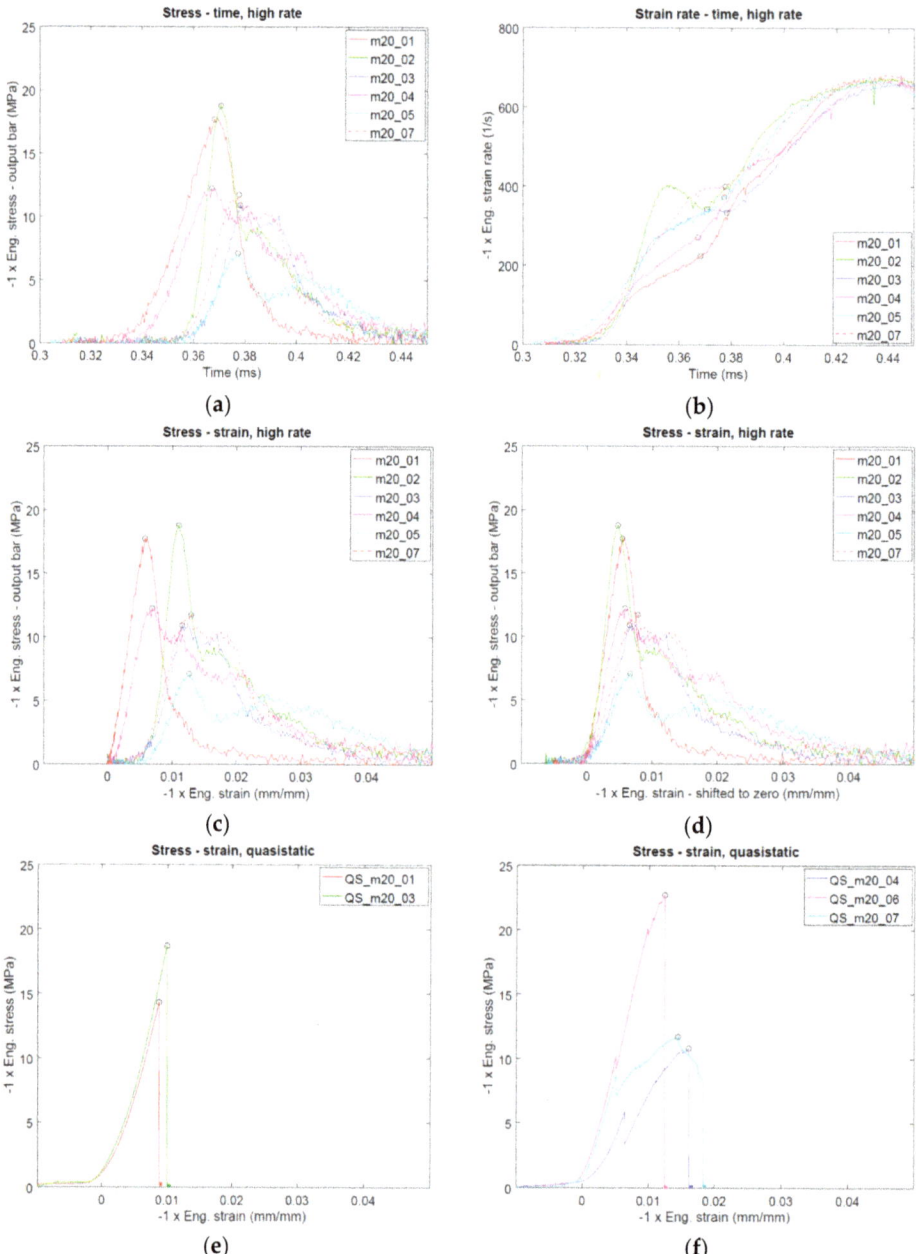

Figure 6. (**a**–**d**): Measured high strain rate data at −20 °C: (**a**) stress-time, (**b**) strain rate-time, (**c**) stress-strain, (**d**) stress-strain with curves shifted to common starting point, (**e**,**f**): measured quasi-static stress-strain data at −20 °C: (**e**) tests with smooth stress-strain curve until maximum stress and (**f**) tests with intermittent drops in stress before maximum stress.

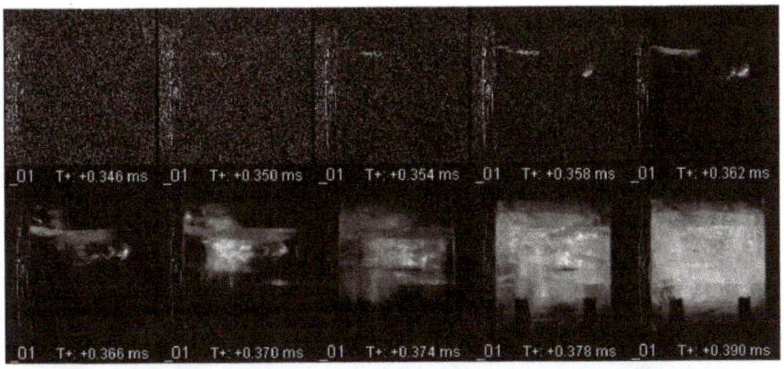

Figure 7. Processed high speed footage obtained in the high rate tests m20_01 (**a**) and m20_05 (**b**) alongside with the corresponding specimen stress–time signal (**c**,**d**) obtained from the output bar strain gauge measurement (synchronized with high speed footage). The input bar is located on the left-hand side of the figure in (**a**,**b**). The time, at which the input bar end velocity exceeded 0.5 m/s as well as the times corresponding to the high speed frames are marked on the respective stress-time curves. The full videos are provided as supplementary material S1 and S2.

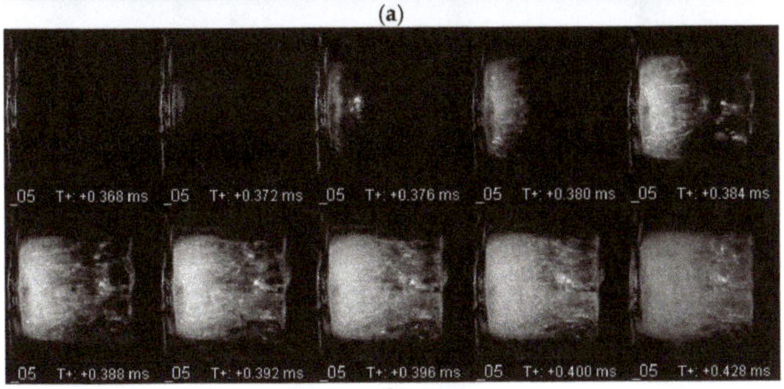

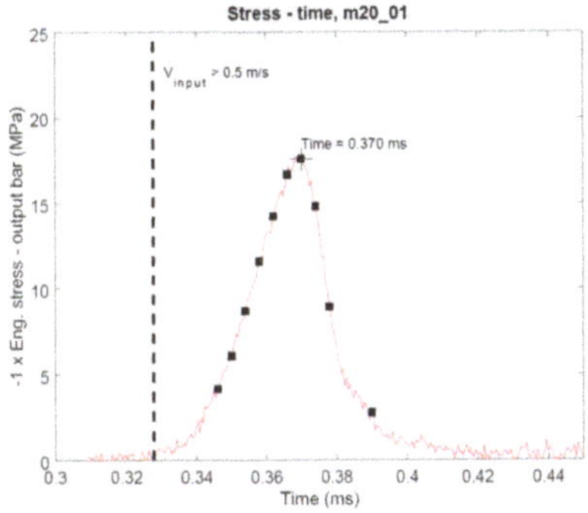

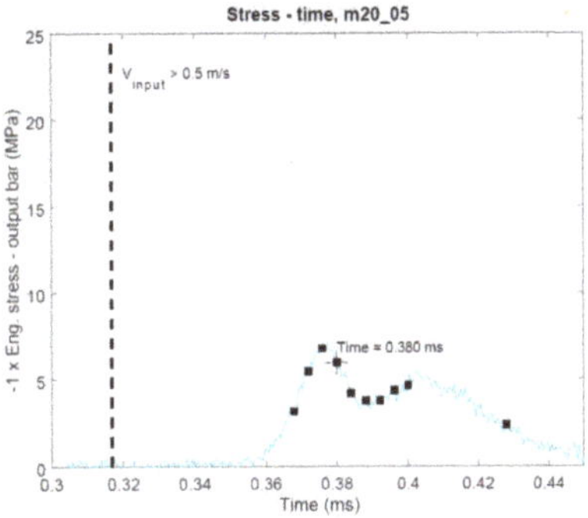

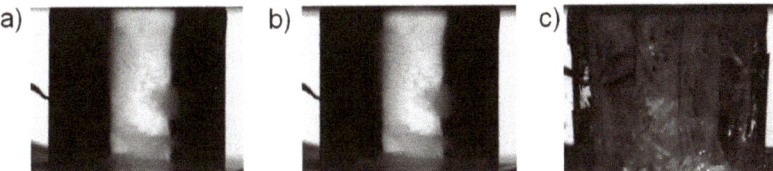

Figure 8. Examples of video frames recorded in the quasi-static test QS_m20_01: (**a**) prior to loading, (**b**) at maximum load, and (**c**) 0.1 s after the previous frame (zero load). The full video is provided as supplementary material S3.

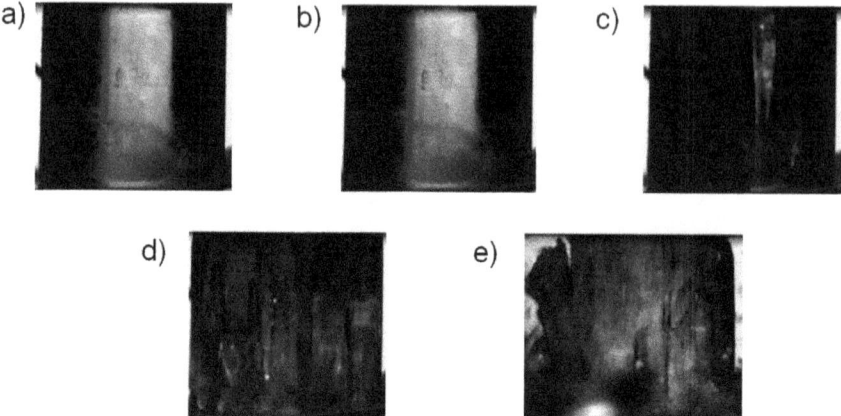

Figure 9. Examples of video frames recorded in the quasi-static test QS_m20_06: (**a**) prior to loading, (**b**) immediately prior to the first drop in load (eng. stress = 10.1 MPa), (**c**) 0.1 s after the previous frame, (**d**) at maximum load (eng. stress 22.7 MPa), and (**e**) 0.1 s after the previous frame (zero load). The full video is provided as supplementary material S4.

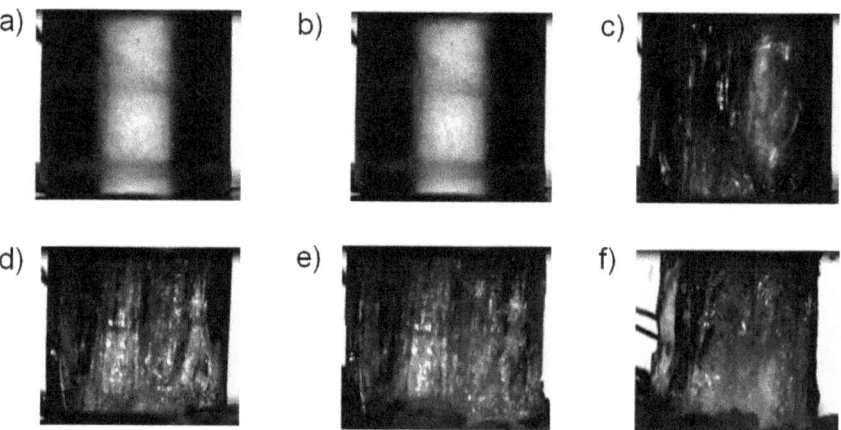

Figure 10. Examples of video frames recorded in the quasi-static test QS_m20_07: (**a**) prior to loading, (**b**) immediately prior to the first drop in load (eng. stress = 8.7 MPa), (**c**) 0.1 s after the previous frame, (**d**) at maximum load (eng. stress = 11.7 MPa), (**e**) after maximum load, immediately prior to the final failure (eng. stress = 8.4 MPa) and (**f**) 0.3 s after the previous frame (zero load). The full video is provided as supplementary material S5.

4. Discussion

As was expected from previous reports (c.f. [11]), the strength of ice showed some scatter, which is typical for brittle materials. However, the macroscopic strength of the tested ice is not notably loading rate sensitive in the studied cases. In contrast, the data indicates a clear loading rate dependence in the damage and failure behavior of the specimens. In quasi-static tests specimen failure was observed to take place either by brittle-like immediate cracking at the maximum stress or by gradual formation of longitudinal cracks resulting in columns, which carried load until maximum stress was reached. In both cases almost no residual load carrying capacity of the specimen after reaching the maximum stress was observed. In contrast, in high rate tests initial specimen damage was localized and often diffuse in appearance. The point of maximum stress was observed to coincide relatively well with the point at which damage was seen throughout the specimen length. Furthermore, all tested high rate specimens showed notable residual load carrying capacity after the maximum stress similarly to previous studies [4,6,7].

Based on the high speed footage it is reasonable to assume that the mechanical behavior observed at high rates is closely related to the crack propagation velocity in ice. Due to the diffuse nature of the observed high rate damage, the velocity of an individual crack is challenging to determine based on the current data. However, an order of-magnitude estimate for the propagation of local damage within the specimen can be obtained by assuming that initial damage forms at one of the contact surfaces of the specimen immediately after the start of loading on the specimen/output bar-interface and that maximum stress is reached when the specimen is damaged throughout its length. This means that the propagation time of the cracks through the specimen was in the same order of magnitude as the overall test duration. With these assumptions and by noting that the time until maximum stress was between 15 µs and 40 µs in the tests, it can be calculated that damage propagates through the specimen (length 20 mm) at an average velocity between 500 m/s and 1300 m/s. Even though this calculation is only a rough estimation, it results in a propagation speed which is in accordance with previous reports. Lian et al. [10] reported a value of 1000 m/s for uniaxial compression tests carried out at the strain rate of $10\ \text{s}^{-1}$. Pereira et. al [12] reported a value of ~2400 m/s in impact tests carried out at ~200 m/s on cylindrical specimens, whereas Tippman et al. [3] reported a value of 2000 m/s for spherical impact specimen traveling at 60 m/s. Furthermore, the crack propagation velocity is in the same order of magnitude as the longitudinal elastic wave speed in ice (the continuum properties, density 897.6 kg/m3 and Young's modulus 9.31 GPa, reported by Carney et al. [2], result in a longitudinal elastic wave speed of 3200 m/s). On the other hand, Smith and Kishoni [13], reported based on ultrasonic measurements elastic wave speeds of 3940 m/s and 1990 m/s for a compressional and a shear wave in ice, respectively. It thus seems that the cracks in ice might be able to propagate close to the elastic shear wave speed.

For the current high strain rate test series the above discussed notion leads to an important conclusion: It is likely that when specimen damage initiates and propagates early on during the loading, a state of full force equilibrium does not necessarily exist in the whole specimen volume, which implies that the stress measured at the output bar interface does not necessarily indicate the stress near the crack: If one assumes a longitudinal elastic wave speed of 3200 m/s, then the aforementioned rise times until maximum stress, 15 µs and 40 µs, result in distances of 48 mm and 128 mm traveled by the longitudinal elastic loading wave, respectively. These distances correspond approximately to 1 and 3 back-and-forth reflections within the 20 mm long test specimen, respectively. This low number of reflections gives reason to suspect that already at this loading rate the specimen may not be in full force equilibrium and that the stress measured at the specimen/output bar-interface does not necessarily indicate the dynamic strength of the material. It seems likely that the measured stress is affected by the elastic unloading waves initiated from the cracks, which form early during the loading and propagate at a high velocity through the specimen volume. Thus, the local stress acting on the material volume, in which a crack is formed, might differ from the one indicated by the measurement at the specimen/output bar –interface.

The second observation of the test series, i.e., the dynamic load carrying capacity of the specimen after maximum stress seems to be explained by the fact that at high rates the specimen still maintains its coherence for a period of time after reaching the maximum stress despite the extensive damage. By noting that in the high rate tests the post-maximum stress period lasted 30 µs to 70 µs, it seems that in the high rate tests the specimen fragments alone are able to hold them together. Thus, the input bar, which is moving at a velocity of ~10 m/s, imparts a compressive load through the fragmented specimen to the output bar. In contrast, in quasi-static loading the maximum stress corresponds to the point at which the specimen loses its coherence and ability to carry quasi-static load. A more detailed analysis of the residual loading capacity would, however, involve incorporating the possible interaction between the particles, such as the presence of liquid water suggested by Wu and Prakash [7], and is considered beyond the scope of the current study.

5. Conclusions

In this work the effect of loading rate on the damage and fracture behavior of pure ice at −20 °C under uniaxial compression was studied. Whilst the ice may not be representative for naturally grown hail, all specimens have been produced with the same methodology, thus allowing qualitative and quantitative assessments within the batch. High rate tests were carried out with the Split Hopkinson Pressure Bar technique in the strain rate region of $100\ s^{-1}$ to $600\ s^{-1}$. Specimen damage and fracture was studied in-situ using high speed imaging. Based on the results the following conclusions can be drawn:

- Under high rate loading damage may initiate and propagate in the specimen even before the peak load is reached. Catastrophic damage only occurs after the peak load was reached.
- Under high rate loading the damage was observed to propagate in the specimen with a rate, which is in the same order of magnitude as the velocity of the elastic loading wave in ice.
- The above-mentioned findings lead to the conclusion that a state of full force equilibrium is not ensured in the specimen when damage initiates and propagates. This finding implies that already at this loading rate the determination of the strength of ice is affected by wave propagation effects. The force equilibrium can theoretically be improved by replacing the aluminum bars with bars matching the impedance of ice more closely. Amongst practical engineering materials, the next option after aluminum would be a technical polymer. However, this approach is believed to bring about more problems than solutions in the current case: The bars would be deforming visco-elastically, thus demanding tedious numerical techniques and experimental calibration in order to accurately interpret the strain gauge signals in the presence of notable dispersion and frequency-dependent damping in the wave motion. In the current study this would be exceptionally challenging due to the short duration of the tests and brittle response of the specimen, which would inevitably lead to notable distortion of the stress waves, as they travel in visco-elastic bars. Furthermore, the mechanical properties of the visco-elastic polymer bars would be most likely affected, when they are subjected to sub-zero temperature in the specimen cooling chamber.
- The fractured specimen can carry notable load when it is compressed at 10 m/s, but not when it is compressed at 0.05 mm/s. This can be explained by dynamic effects which do not occur under quasi-static loading conditions which are four orders of magnitude smaller than the high-rate loading conditions achieved using the Hopkinson Bar apparatus.
- For quasi-static loading, the measured engineering mean peak stress was 15.6 MPa (COV 31.6%), for high-rate loading, the measured engineering mean peak stress was 12.4 MPa (COV 34.1%) indicating a decrease of strength with increasing loading rate. However, at the same time, the post failure response changes with loading rate such that the post-peak load carrying capability is higher for high loading rates. However, due to the small sample size, this lacks stochastic relevance and should be assessed with more tests.

The results of this study clearly show the importance of using in-situ high speed imaging, when the high rate response of ice is studied. A variety of initial cracking morphologies was observed. By using only macroscopic stress-strain curves without in-situ footage, these differences in specimen behavior would be very challenging to detect. This fact has clear implications for example in the development and calibration of high strain rate material models for ice, whose accuracy in many cases relies on the correct description of the fracture propagation within the material.

Supplementary Materials: The following are available online at https://zenodo.org/record/2649030#.XL7aX5MRWUk, Video S1: Video recording of Hopkinson Bar test HR_m20_01 (input bar on the left, output bar on the right), Video S2: Video recording of Hopkinson Bar test HR_m20_05 (input bar on the left, output bar on the right), Video S3: Video recording of quasi-static test QS_m20_01. Video S4: Video recording of quasi-static test QS_m20_06. Video S5: Video recording of quasi-static test QS_m20_07.

Author Contributions: Conceptualization, M.I. and M.M.; Methodology, J.L., S.K. and M.I.; Formal Analysis, M.I.; Investigation, M.I., J.L.; Writing—Original Draft Preparation, M.I.; Writing—Review & Editing, M.M., M.I., S.K.; Supervision, M.M., S.K.; Funding Acquisition, M.M.

Funding: The funding for the experimental campaign was provided by the European Union funded Clean Sky 2 project (Grant Agreement no. CS2-AIR-GAM-2016-2017-05).

Conflicts of Interest: The authors declare no conflict of interest.

Appendix A

This appendix shows additional high-speed footage recorded during SHPB testing.

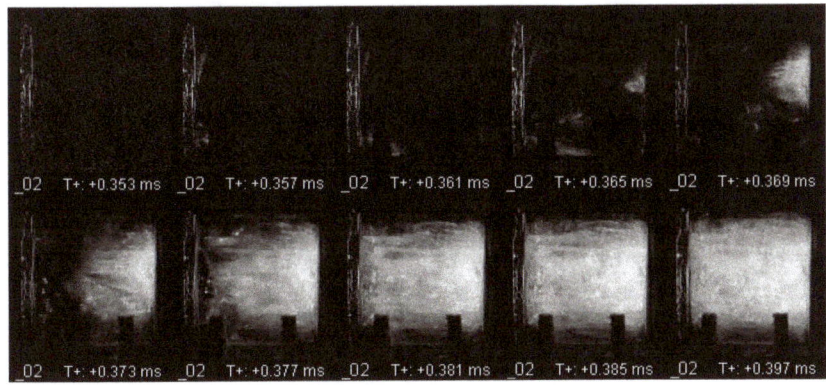

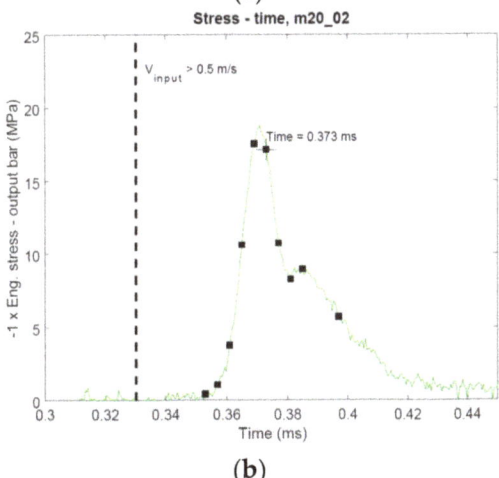

Figure A1. (**a**) processed high speed footage obtained in the high rate test m20_02, (**b**) corresponding specimen stress-time signal obtained from the output bar strain gauge measurement (synchronized with high speed footage). The input bar is located on the left-hand side of the figure in (**a**). The time, at which the input bar end velocity exceeded 0.5 m/s as well as the times corresponding to the high speed frames are marked on the respective stress-time curve.

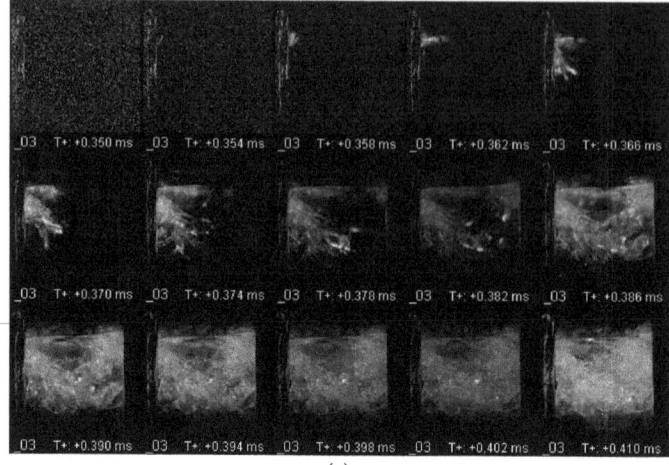

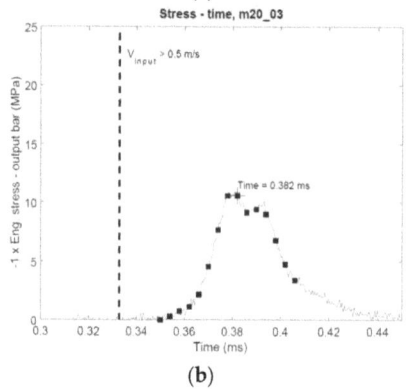

Figure A2. (**a**) Processed high speed footage obtained in the high rate test m20_03, (**b**) corresponding specimen stress-time signal obtained from the output bar strain gauge measurement (synchronized with high speed footage). The input bar is located on the left-hand side of the figure in (**a**). The time, at which the input bar end velocity exceeded 0.5 m/s as well as the times corresponding to the high speed frames are marked on the respective stress-time curve.

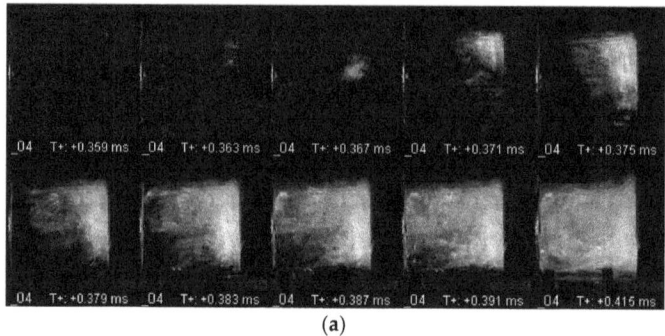

(**a**)

Figure A3. *Cont.*

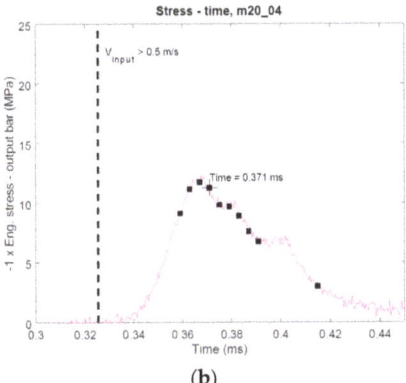

(b)

Figure A3. (**a**) processed high speed footage obtained in the high rate test m20_04, (**b**) corresponding specimen stress-time signal obtained from the output bar strain gauge measurement (synchronized with high speed footage). The input bar is located on the left-hand side of the figure in (**a**). The time, at which the input bar end velocity exceeded 0.5 m/s as well as the times corresponding to the high speed frames are marked on the respective stress-time curve.

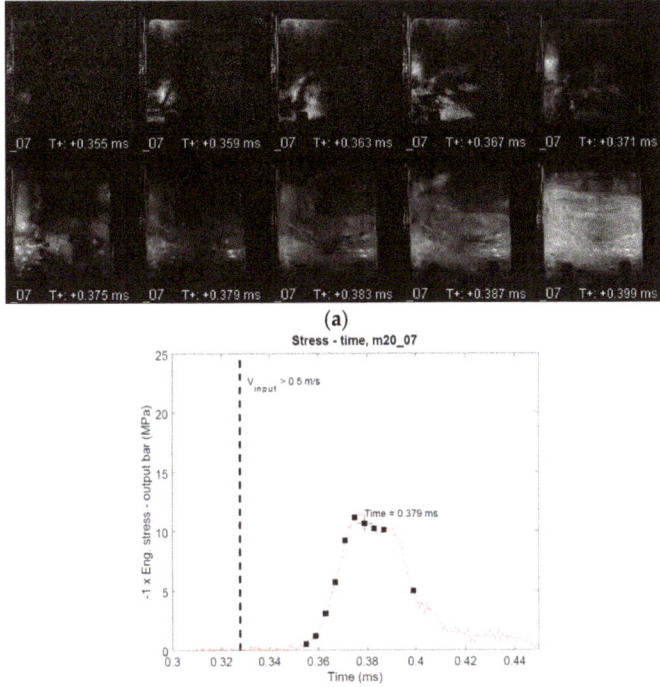

Figure A4. (**a**) processed high speed footage obtained in the high rate test m20_07, (**b**) corresponding specimen stress-time signal obtained from the output bar strain gauge measurement (synchronized with high speed footage). The input bar is located on the left-hand side of the figure in (**a**). The time, at which the input bar end velocity exceeded 0.5 m/s as well as the times corresponding to the high speed frames are marked on the respective stress-time curve.

References

1. *Hail Threat Standardization FINAL Report for EASA.2009.OP.25*; European Aviation Safety Agency: Cologne, Germany, 2010.
2. Carney, K.S.; Benson, D.J.; DuBois, P.M.; Lee, R. A phenomenological high strain rate model with failure for ice. *Int. J. Solids Struct.* **2006**, *43*, 7820–7839. [CrossRef]
3. Tippman, J.D.; Kim, H.; Rhymer, J.D. Experimentally validated strain rate dependent material model for spherical ice impact simulation. *Int. J. Impact Eng.* **2013**, *57*, 43–54. [CrossRef]
4. Shazly, M.; Prakash, V.; Lerch, B.A. High strain-rate behavior of ice under uniaxial compression. *Int. J. Solids Struct.* **2009**, *46*, 1499–1515. [CrossRef]
5. Kim, H.; Keune, J.N. Compressive strength of ice at impact strain rates. *J. Mater. Sci.* **2007**, *42*, 2802–2806. [CrossRef]
6. Bragov, A.; Igumnov, L.; Konstantinov, A.; Lomunov, A.; Filippov, A.; Shmotin, Y.; Didenko, R.; Krundaeva, A. Investigation of strength properties of freshwater ice. *EPJ Web Conf.* **2015**, *94*, 01070. [CrossRef]
7. Wu, X.; Prakash, V. Dynamic compressive behavior of ice at cryogenic temperatures. *Cold Reg. Sci. Technol.* **2015**, *118*, 1–13. [CrossRef]
8. Song, Z.; Wang, Z.; Kim, H.; Ma, H. Pulse Shaper and Dynamic Compressive Property Investigation on Ice Using a Large-Sized Modified Split Hopkinson Pressure Bar. *Lat. Am. J. Solids Stru.* **2016**, *13*, 391–406. [CrossRef]
9. Chen, W.; Song, B. *Split Hopkinson (Kolsky) Bar – Design, Testing and Applications*; Springer Science+Business Media: New York, NY, USA, 2011.
10. Lian, J.; Ouyang, Q.; Zhao, X.; Liu, F.; Qi, C. Uniaxial Compressive Strength and Fracture of Lake Ice at Moderate Strain Rates Based on a Digital Speckle Correlation Method for Deformation Measurements. *Appl. Sci.* **2017**, *7*, 495. [CrossRef]
11. Petrovic, J.J. Review Mechanical properties of ice and snow Compressive strength of ice at impact strain rates. *J. Mater. Sci.* **2003**, *38*, 1–6. [CrossRef]
12. Pereira, J.M.; Padula, S.A.; Revilock, D.M.; Melis, M.E. *Forces Generated by High Velocity Impact of ice on a Rigid Structure*; BiblioGov: London, UK, 2013.
13. Smith, A.C.; Kishoni, D. Measurement of the Speed of Sound in Ice. *AIAA J.* **1986**, *24*, 1713–1715. [CrossRef]

© 2019 by the authors. Licensee MDPI, Basel, Switzerland. This article is an open access article distributed under the terms and conditions of the Creative Commons Attribution (CC BY) license (http://creativecommons.org/licenses/by/4.0/).

Article

Evolution of Thermal Microcracking in Refractory ZrO₂-SiO₂ after Application of External Loads at High Temperatures

René Laquai [1], Fanny Gouraud [2], Bernd Randolf Müller [1,*], Marc Huger [2], Thierry Chotard [2], Guy Antou [2] and Giovanni Bruno [1,3]

[1] Bundesanstalt für Materialforschung und-prüfung (BAM), Unter den Eichen 87, D-12200 Berlin, Germany; rene.laquai@bam.de (R.L.); giovanni.bruno@bam.de (G.B.)
[2] Centre Européen de la Céramique, University of Limoges, 12 Rue Atlantis, 87068 Limoges, France; fanny.gouraud@gmail.com (F.G.); marc.huger@unilim.fr (M.H.); thierry.chotard@unilim.fr (T.C.); guy.antou@unilim.fr (G.A.)
[3] Institute of Physics and Astronomy, University of Potsdam, Karl-Liebknecht-Str.24-25, 141176 Potsdam, Germany
* Correspondence: bernd.mueller@bam.de; Tel.: +49-30-8104-1852

Received: 25 February 2019; Accepted: 22 March 2019; Published: 27 March 2019

Abstract: Zirconia-based cast refractories are widely used for glass furnace applications. Since they have to withstand harsh chemical as well as thermo-mechanical environments, internal stresses and microcracking are often present in such materials under operating conditions (sometimes in excess of 1700 °C). We studied the evolution of thermal (CTE) and mechanical (Young's modulus) properties as a function of temperature in a fused-cast refractory containing 94 wt.% of monoclinic ZrO_2 and 6 wt.% of a silicate glassy phase. With the aid of X-ray refraction techniques (yielding the internal specific surface in materials), we also monitored the evolution of microcracking as a function of thermal cycles (crossing the martensitic phase transformation around 1000 °C) under externally applied stress. We found that external compressive stress leads to a strong decrease of the internal surface per unit volume, but a tensile load has a similar (though not so strong) effect. In agreement with existing literature on β-eucryptite microcracked ceramics, we could explain these phenomena by microcrack closure in the load direction in the compression case, and by microcrack propagation (rather than microcrack nucleation) under tensile conditions.

Keywords: electro-fused zirconia; microcracking; synchrotron x-ray refraction radiography (SXRR); thermal expansion

1. Introduction

The manufacturing of high quality glasses required for new applications (e.g., flat LCD or PDP screens) imply the development of new, high zirconia fused-cast refractories with excellent thermomechanical properties [1]. In order to build suitable refractory linings for glass furnaces, this 'high zirconia' (meaning high zirconia content) material is typically cast in heavy prismatic industrial blocks (about 1 m³) that are then adjusted to build the inner wall (lining). During the controlled cooling step after casting at about 2500 °C, dendrites of zirconia initially grow under the form of cubic domains (C), but then transform into tetragonal domains (T) at around 2300 °C. Upon further cooling, between 1000 °C and 900 °C, zirconia transforms from a tetragonal to a monoclinic (M) crystal structure. This transformation is associated with a large volume expansion of about 5% [2]. This generates large local (micro) stresses and typically microcracks [3]. While the polymorphism in zirconium ceramics is very well known [4,5], most of the literature is limited to the case of structural ceramics utilized in a rather low temperature range. In the present case, a high zirconia refractory material is targeted to very high temperature applications (typically 1500 °C, and even in excess of 1700 °C) within industrial furnaces for the production of 'special' glasses. In such a case, the stabilized (or partially stabilized) zirconia would be inadequate, since the refractory material is used above the zirconia phase transition (taking place between 1000 °C and 1160 °C in pure zirconia). Instead, a small quantity of glassy phase (SiO_2) within the refractory microstructure is introduced to facilitate the industrial processing of large blocks (not otherwise possible). The glassy phase accommodates internal stresses induced by the large volume expansion associated to the zirconia phase transition during cooling (and also during preheating to the service temperature in the industrial furnace).

During the cooling stage, refractory blocks are also submitted to thermal gradients between their core and their skin. Thereby the martensitic transformation of zirconia usually occurs under a significant thermal gradient that generates additional macro-stresses, which can further significantly modify local microcracking mechanisms and induce macrocracks.

Thermally-induced microcracking (in the following simply thermal microcracking) generally occurs in ceramics with anisotropic coefficient of thermal expansion (CTE) when cooling from the sintering temperature [6]. Thermal microcracking distinguishes itself from mechanical microcracking, since it is: (a) Reversible and (b) mostly insensitive to the thermal cycle history [6,7]. Typical materials undergoing thermal microcracking are aluminum titanate [7,8], cordierite [8] and β-eucryptite [9,10]. The same phenomenon can occur in composites, where the (brittle) constituents have different CTE. This phenomenon is visible in the hysteresis of both the thermal dilatation (and of the CTE) and of the Young's modulus as a function of temperature [11]. In the present study, high zirconia fused-cast refractory are typically constituted by pure ZrO_2 embedded in a silica-rich glassy phase (though present in very small amounts). In this case, the network of microcracks stems from the 5% volume expansion associated to the T → M transformation in the 1000 °C to 900 °C range, but also from the anisotropic thermal expansion coefficient of the monoclinic (M) phase below 1000 °C. This network can be significantly modified by the additional macro-stress field induced by the thermal gradient during the whole cooling process. Thus, on one hand, industrial cooling condition of the blocks could greatly influence local microcracking, and on the other hand this network of microcracks could also significantly modify the thermo-mechanical properties (CTE, thermal conductivity, elastic properties). Depending on the microstructural features (dendrites and domains size), microcracks could play a key role in the appearance of macrocracks, which would render the block unusable. This is why a better understanding of the evolution of microcracking under operating conditions (especially thermal cycles, as well as mechanical loads) is of particular interest for both fundamental and industrial aspects. Due to the inhomogeneity of thermal stresses (large differences between core and periphery), we needed to reproduce the effect of the local stress field undergone by the materials during annealing and, more particularly, during the phase transition. Therefore, we reported that specimens were submitted to different levels of uniaxial load (tensile or compressive) during the phase transition (T → M) of zirconia. In this way, the occurrence of a TRansformation Induced Plasticity (TRIP) phenomenon during the zirconia phase transition could be characterized (such phenomenon is extremely poorly documented in the literature [12,13]). Since one current hypothesis is that the applied stress generates additional damage through microcracking, and both density and orientation of such microcracks would influence the 'plastic' deformation of the transition, the main target of the present work was to monitor the effect of thermal history (thermal cycling with/without external uniaxial loading) on the network of microcracks and on some thermomechanical properties (since a uniaxial load potentially leads to some anisotropy of such properties).

With this aim, we used novel techniques such as X-ray refraction radiography. The application of X-ray refraction [14] as a microstructure characterization technique at a macroscopic scale has proven to provide answers to questions that cannot be tackled even by the highest resolution techniques such as computed tomography. This is because X-ray refraction techniques possess detectability of features (cracks, pores, etc.) with size (~1 nm) well below their spatial resolution (~1 µm in the best case) as well as that of computed tomography. This detection power has been exploited in previous works using X-ray refraction [15,16], whereas model has been elaborated to rationalize the evolution of a network of microcracks in terms of propagation of large microcracks that could lead to the closure of smaller ones.

We will see below that microcracks evolved as a function of mechanical load, and are retained at the end of the thermal cycles. Such microcracks also engender a change of the equivalent elastic constants, therefore impacting the materials properties.

It belongs to future work to embed this damage behavior in new elasto-plastic constitutive laws.

2. Materials and Methods

2.1. Specimen Manufacture

The studied material was a fused-cast refractory containing 94 wt.% of monoclinic ZrO_2 and 6 wt.% of a silicate glassy phase. Materials were typically obtained under the form of large blocks (1 m^3) by a melting process in arc furnaces followed by controlled cooling in ceramic or graphite molds. All investigated specimens were machined by diamond tools from these cast blocks.

For investigations under a uniaxial load at a high temperature, specimens were prepared from cylindrical rods of 20 mm in diameter with metallic parts glued at each end. The final dog bone geometry was obtained by machining simultaneously the central zone (25 mm gauge length) of the specimen down to a diameter of 16 mm and the metallic parts. This allowed obtaining an optimized alignment with the loading axis.

In the second step, for X-ray refraction investigations, smaller prismatic specimens (30 mm × 16 mm × 0.5 mm) were machined in the central zone of the previous ones with a good identification of the load axis. For elastic properties measured by ultrasound in different directions at room temperature, cubic specimens (10 mm × 10 mm × 10 mm) cut with the same procedure, were used.

2.2. Microstructural Characterization

In order to investigate the complex microstructure of these high zirconia fused-cast refractories, coupons were cut and polished for SEM imaging. A Carl Zeiss AG-SUPRA 40 (Carl Zeiss AG, Oberkochen, Germany) with the following experimental conditions was used: Accelerating voltage = 15 kV and specimen-detector distance = 7 mm. A total of 4 material conditions were investigated: As-cast, after purely thermal cycling and after thermal cycling under tensile and compressive loads.

2.3. Dilatometry

A horizontal dilatometer (Netzsch DIL 402 C, NETZSCH-GERÄTEBAU GMBH, Selb, Germany) was used for thermal expansion analysis (no applied load). A large specimen (25 mm length) was prepared in order to reduce measurement error. As usual for such dilatometric experiments, a first test had to be performed on a calibration specimen (high purity sintered alumina), so as to determine and then subtract the specimen holder's dilatation for further experiments on investigated materials.

2.4. Constrained Expansion Measurements

To reproduce the stress field undergone by the materials during the casting process (and during the phase transition), some specimens were submitted to different levels of uniaxial stress when undergoing the T → M transformation. These tests were performed with an INSTRON 8862 electro-mechanical universal testing machine (INSTRON, Norwood, MA, USA), which can operate up to 1600 °C. Strain was measured from the variation of a 25 mm gauge length obtained by 2 capacitive extensometers placed on the opposite faces of the specimen. The specimens were first heated up to 1500 °C with a rate of 10 °C/min, and then dwelled for 1 hour. This allowed them to return to a stress-free state. Then specimens were cooled with different temperature ramps, simulating the industrial production process of these materials. During cooling, the load was applied after a short dwell at 1150 °C. This ensured starting the application of the load at a low enough temperature to limit the risk of rupture, yet well above the T → M transformation. Once the load was applied, it was kept constant down to room temperature. A total of 2 tests were carried out with different stress levels (tension, +1 MPa and compression, −5 MPa). The values of the load were chosen, taking into account the strength values observed in tension and in compression at the temperature of the zirconia phase transition (about 1000 °C). We aimed at maximizing the effect of the TRIP phenomenon on the associated macroscopic deformation. In fact, at above 1.5 MPa tensile, rupture was systematically observed during the zirconia phase transition. The value of −5 MPa compressive was selected so that a more clearly visible effect on the deformation associated with the phase transition could be observed.

A reference cycle without any load (0 MPa) was run. As a side remark, we observed that the free dilation measured with the dilatometer and with the test rig, while having a similar trend, did not quantitatively match. This is due to the different calibrations of the 2 machines.

2.5. Ultrasonic Transmission

Ultrasonic measurements, in different directions, were carried out on specimens in the 4 conditions mentioned above. The velocities of longitudinal waves in infinite mode were measured according to the pulse transmission echo method [17]. The transmission method was applied due to the presence of some defects after thermal cycling under load, which can disturb the signal leading to a false measurement of the velocity if the reflection method were used.

A pulse generator/receiver system and 2 piezoelectric transducers (10 MHz) were used. A transducer was applied on 1 face of the specimen and directly sent ultrasonic waves across it. A second transducer collected the waves on the other face. The received signal was recorded on a digital oscilloscope. After signal analysis, the transit time through the thickness of the specimen was measured and related to wave velocity along a particular direction. Measurements were made in 3 perpendicular directions. The knowledge of the density allowed the calculation of the corresponding elastic constants.

2.6. X-ray Refraction

2.6.1. Generalities about X-ray Refraction Techniques

X-ray refraction techniques were introduced a couple of decades ago [18] and have been successfully used for both material characterization and non-destructive testing [19]. X-ray refraction techniques are used to obtain the amount of the relative internal specific surface (i.e., surface per unit volume, relative to a reference state) of a specimen, and are therefore beneficial in the investigation of defects such as cracks and pores within ceramic components.

X-ray refraction occurs whenever X-rays interact with interfaces between materials of different densities as in the case of cracks, pores and particles in a matrix. The difference in the refraction indices between the 2 interfacing materials, the so-called refraction decrement, determines the refraction angle at the interface. The refraction decrement is dependent on the wavelength of the radiation. Since the refraction decrement for X-ray radiation is of the order of 10^{-5}, X-ray optical effects can only be observed at very small scattering angles, which lie between a few seconds and a few minutes of arc. Since the typical X-ray wavelengths are approximately 0.1 nm, X-ray refraction detects pores and cracks as soon as they exceed a size of some X-ray wavelengths (so that the wave 'notices' the density difference at the interface). That means the smallest detectable object size is down to the nanometer range. This is not to be confused with the spatial resolution or the size of the objects that can be imaged. The spatial resolution is limited by the pixel size (in this work of about 4 μm × 4 μm) of the detector system. It must be emphasized that because of the inevitable background noise, it is impossible to conclusively detect one single defect. A certain population of objects is necessary to yield an integrated signal above the background noise. Thus, X-ray refraction is used primarily in radiographic mode with thin specimens (platelets) and yields a 2.5D signal, i.e., integrated over the specimen's thickness. This results in the detection and imaging of a population of defects rather than the imaging of single defects. The X-ray refraction signal has been quantitatively correlated to microstructural changes and micromechanical models [20]. Furthermore, X-ray refraction techniques are sensitive to defect orientation, thereby allowing different kinds of defects to be identified [21]. The refraction signal of an isotropic inhomogeneity, such as spherical voids, will also be isotropic, whereas for cracks or elongated pores the signal vanishes when the defect surface normal is oriented perpendicular to the scattering vector of the detection system.

2.6.2. Synchrotron Radiation X-ray Refraction

SXRR measurements were carried out at the BAM synchrotron laboratory BAMline at Helmholtz-Zentrum, Berlin, Germany [16,22]. The 3 specimens were mounted in a slide frame as shown in Figure 1. The beam energy was set to 50 keV (with $\Delta E/E$ ~0.2%) to achieve a specimen X-ray transmission of about 30%. A Princeton Instrument camera (2048 × 2048 pixel) in combination with a lens system and a 50 µm thick CWO scintillator screen provided a pixel size of 3.5 µm × 3.5 µm, capturing a field of view of about 7 mm × 3 mm [23].

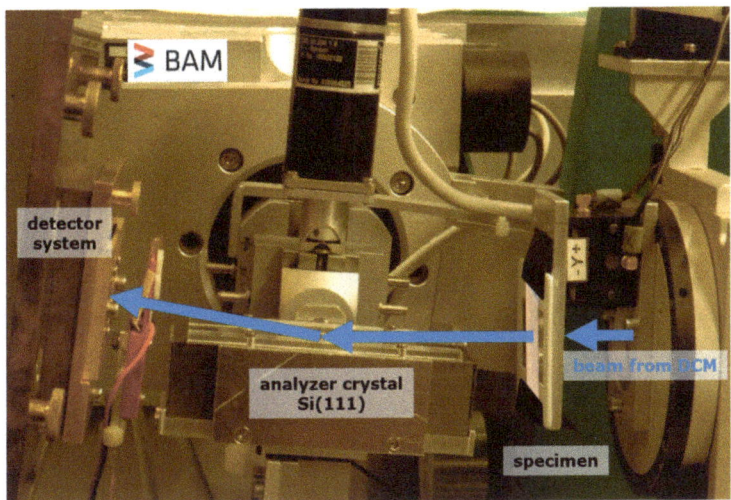

Figure 1. Experimental set up of the X-ray refraction station at BAMline. The specimens were mounted in a slide frame shown on the right. The scattering plane of the Si(111) crystal lies in the image plane.

A Si(111) analyzer crystal was placed in the beam path between the specimen and the camera system, shown in Figure 1, to perform refraction radiographs. The analyzer crystal reflect the beam coming out of the specimen into the detector system if the incidence angle is set to the Bragg angle, $\theta_B = 2.2664°$ at 50 keV. By tilting the analyzer crystal around an axis perpendicular to the scattering plane (the scattering plane of the Si(111) crystal lies in the plane of Figure 1), the so-called rocking curve is recorded. This describes the reflected beam intensity as a function of the deviation from the Bragg angle, $\Delta\theta = \theta - \theta_B$. The rocking curve was recorded for each specimen by taking 41 radiographs between $\theta = 2.2651°$ and $\theta = 2.2685°$ with a step size of $\Delta\theta = 0.0001°$ and exposure time of 5 s. All specimens were measured in two orientations: (a) Load direction of the specimen perpendicular and (b) parallel to the scattering plane (see Figure 1). In addition, the following images were acquired: Dark field (beam off) and flat field (beam on, but without specimens). The dark field image was used to subtract the dark current and detector readout noise from the specimen and flat field acquisitions. The flat field images quantified the instrumental artefacts and noise that were used to correct the X-ray radiography images according to Equation (1). The corrected rocking curve images were analyzed using an in-house software code based on LabView®. Figure 2 shows the typical rocking curves extracted from one arbitrary detector pixel. Open circles indicate the measurement without and filled circles with the specimen in the beam, respectively. The peak maxima are normed to unit to show the increase in beam divergence due to the refraction effect at interfaces (e.g., cracks and/or pores) inside the specimen. The analysis software delivered the values of the rocking curve integral, the peak height, peak position and the Full Width at Half Maximum FWHM (see Table 1). By using the image calculating software "Fiji Image J" [24] the attenuation ($\mu \cdot d$) and the refraction value ($C_m \cdot d$) were evaluated for each pixel according to Equations (1) and (2), respectively. A detailed description of the data processing and evaluation can be found in [25,26].

$$\mu \cdot d = \ln\left(\frac{I_0}{I}\right) \quad (1)$$

$$C_m \cdot d = 1 - \frac{I_R \cdot I_0}{I_{R0} \cdot I} \quad (2)$$

The influence of the specimens' thickness d is eliminated by dividing the local refraction value ($C_m \cdot d$) by the local attenuation property ($\mu \cdot d$). This yielded the relative specific refraction value (C_m/μ), which is proportional to the relative specific internal surface of the specimen up to a calibration constant, depending on the instrument, the material and the experiment (geometry and energy).

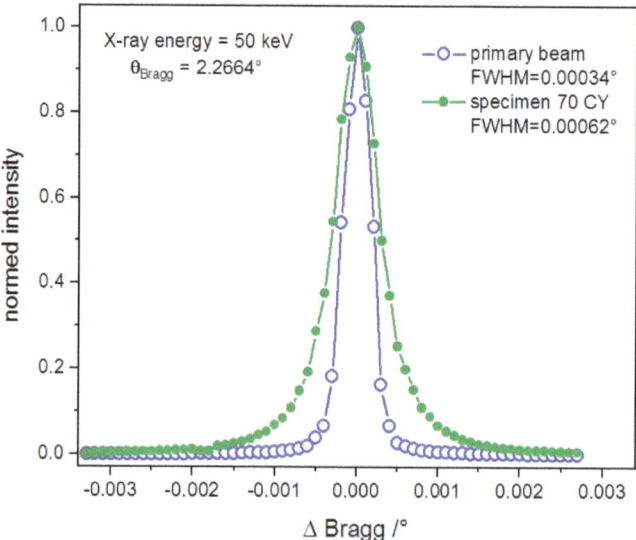

Figure 2. Rocking curves measured at one detector pixel and normalized to the peak maximum. Open circles: Without specimen (FWHM = 0.00034°); and filled circles: With specimen (FWHM = 0.00062°).

Table 1. The rocking curve parameter used to calculate the attenuation properties and the relative specific refraction value of the specimens.

Symbol	Quantity
I_R	peak height (curve with filled circles) with specimen in the beam
I_{R0}	peak height (curve with open circles) without specimen in the beam
I	peak integral (curve with filled circles) with specimen in the beam
I_0	peak integral (curve with open circles) without specimen in the beam

3. Results

3.1. Microstructure

Microstructural features associated with the cooling process are very complex. Figure 3a illustrates the different steps of microstructure evolution during the different stages of the cooling process [27]. At a very high temperature (2500 °C), dendrites of zirconia initially grew with a cubic structure (C). These dendrites possess primary and secondary ramifications (tree structure in Figure 3c); the structure transformed into tetragonal domains (T) at around 2300 °C. Between 2300 °C and 1700 °C, the mix (Figure 3b) was not supposed to be fully solid, and nucleation-growth of zirconia dendrites probably continued in this temperature range. Below 1700 °C, the material could be considered as fully solid with zirconia dendrites embedded in a silica glassy phase (Figure 3d). Between 1000 °C and 900 °C, the martensitic transformation of zirconia from the tetragonal to the monoclinic structure (M) occurred (Figure 3a). The cubic-to-tetragonal transformation was associated to a 45° rotation of the a- and b-crystal axes around the c-axis. This rotation induced the possible formation of three distinct crystallographic variants from one single cubic crystal. During the tetragonal-to-monoclinic transformation, it was possible to form 24 different crystallographic variants. The β angle (between a and b) differed thus from 90° (being close to 99°). At room temperature each zirconia dendrite was therefore constituted of different monoclinic variants (Figure 3e,f). Considering the anisotropy in thermal expansion along the different crystallographic axes in the monoclinic structure, these different crystallographic variants induced thermal mismatches and then potential microcracking between variants (Figure 3f). The glassy phase within the microstructure was assumed, because of its low viscosity at this temperature, to accommodate internal stresses induced by the anisotropic expansion mismatch between ZrO_2 grains during this transformation.

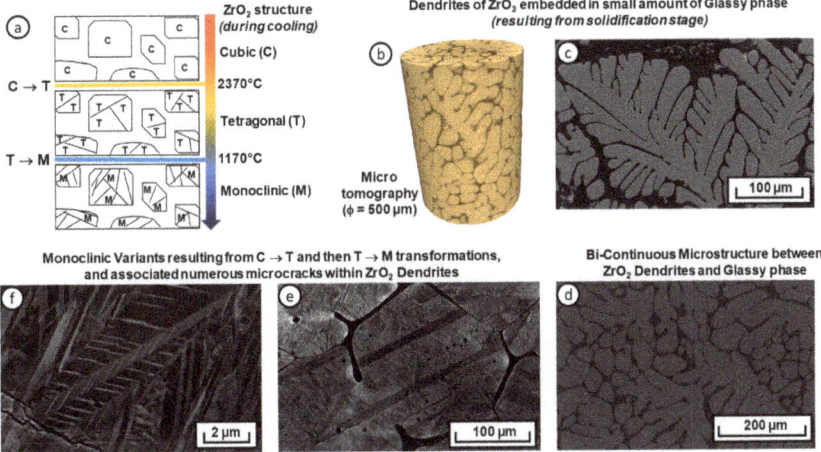

Figure 3. (a) Microstructure evolution of the ZrO_2-SiO_2 refractory; (b) 3D microtomography reconstruction of a cylindrical specimen, showing the dendritic structure, as in (c,d) (SEM pictures). (e,f) show the monoclinic variants stemming from the C → T and then T → M phase transformations.

3.2. Free and Constrained Thermal Expansion

As a reference without any external applied stress, a classic dilatometric experiment was carried out up to 1500 °C. This determined the characteristic temperatures for phase transformations of zirconia, as well as the average amplitude of the different effects in the unconstrained case (Figure 4). The general shape of this curve described an open hysteresis cycle with expansion discontinuities due to dimensional changes in the zirconia structure associated with transformations M ↔ T. The M → T transformation took place around 1115 °C, during heating (beginning of curve descent), whereas the inverse transformation occurred around 1010 °C, during cooling (beginning of sudden expansion). The amplitude of linear expansion associated with the phase transformations of zirconia depend on the zirconia content. In the present case, an expansion of 1.7% during cooling corresponded to a volume variation of 5.1%. This experimental value on high zirconia fused-cast refractories can be compared with the intrinsic volume variation associated with the transformation T → M in the case of a pure zirconia that can be measured by X-ray diffraction at high temperatures (3.6%) [12,28]. The macroscopic linear expansion of the present material was much larger than the microscopic (lattice) one, measured by XRD. The presence of a small amount of vitreous (silica) phase (6 wt.%, i.e., 12 vol.%), which has a slightly lower thermal expansion than zirconia, cannot explain this difference. This difference can only be explained by the variation of free space (void, microcracks) within the microstructure.

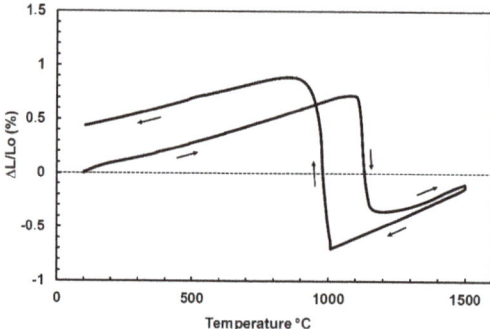

Figure 4. Unconstrained thermal expansion curve of the ZrO_2-SiO_2 refractory.

Results of the constrained dilation tests are presented in Figure 5, where strains are presented using 1500 °C as the reference state. It is clearly observed that the applied stress directly affected the strain associated with the tetragonal to monoclinic transformation. A tensile stress increased the strain associated to the T → M transformation, whereas a compressive one reduced it.

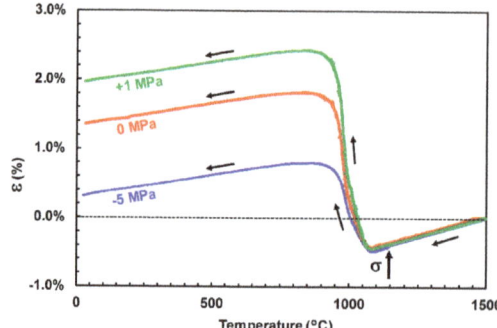

Figure 5. Constrained thermal expansion curve of the ZrO_2-SiO_2 refractory upon cooling, calculated assuming the strain-free state to hold at the maximum temperature (1500 °C). The stress was applied just before the start of the phase transformation (see vertical arrow).

The ability of an external stress to influence the deformation associated with a phase transition has been already observed in metals [29]. This phenomenon is known as Transformation Induced Plasticity. Thus, an interpretation of the present results can be based on similar mechanisms. The TRIP is a permanent macroscopic deformation occurring in the materials subjected to a phase transformation under an external mechanical stress even if this stress is much lower than the yield limit of the different phases that are present in the material. From a microstructural point of view, two mechanisms are usually considered to explain TRIP phenomenon in metals:

- The Greenwood–Johnson mechanism [30] corresponds to the micro-plasticity at the grain boundaries, which is required for the accommodation of the density differences during the phase transformation;
- The Magee mechanism [31] corresponds to a selective orientation of some crystallographic variants depending on the direction of the applied stress.

In the case of brittle material such as zirconia, a third mechanism corresponding to damage is also involved: Microcracking. The strain associated to microcracking is related thus to the number, the width and the orientation of microcracks generated under stress. The evolution of the microcrack density can be quantified by an elastic property measurement [8,9]. To this aim, Young's modulus of all specimens was measured at room temperature before and after each test, applying a compressive stress between 0 MPa and 0.5 MPa. The relative changes of Young's modulus after different cycles are reported below. A decrease in Young's modulus was systematically observed after each test. In addition, this decrease was strongly correlated with the applied stress, being larger in the case of thermal cycles run under tensile stress. A thermal cycle run under compression induced a smaller decrease in Young's modulus in comparison with an unconstrained thermal cycle.

3.3. Microcracking

Figure 6 shows the local linear attenuation coefficient μ as 2D color-coded images (the color spread is the same for all images). The 0 MPa and +1 MPa specimens show similar values of the linear attenuation coefficient. The attenuation of the 5 MPa specimens is about 5% higher. The attenuation is not homogeneous (areas of higher attenuation intersect with a network of lower attenuation). The spatial distribution of the attenuation is similar for all specimens, but the peak attenuation is higher for the 5 MPa specimen.

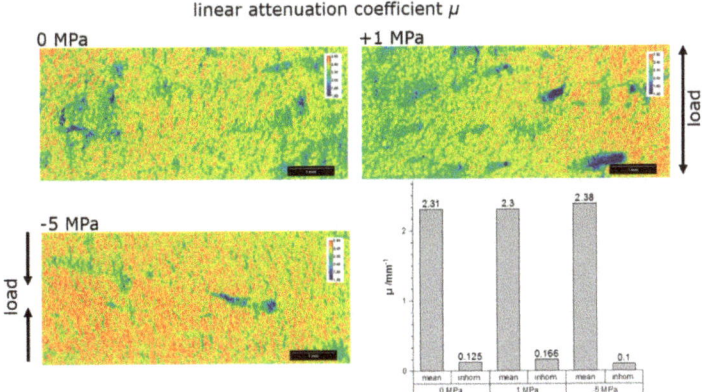

Figure 6. Visualization of the local values of the linear attenuation coefficient μ as 2D color-coded images (top left), unloaded (top right) 1 MPa tensile loaded and (bottom left) 5 MPa compression loaded. Bottom right: The integral values of the linear attenuation coefficient μ across the loaded and unloaded specimens are shown as bar graphs. The small bars represent the inhomogeneity of the μ values across the measured area of the specimen.

Figures 7 and 8 show the local relative specific refraction value C_m/μ as 2D color-coded images, respective for the load direction of the specimen perpendicular- and parallel-oriented to the scattering plane of the analyzer crystal (the color spread is the same for all images) for each specimen.

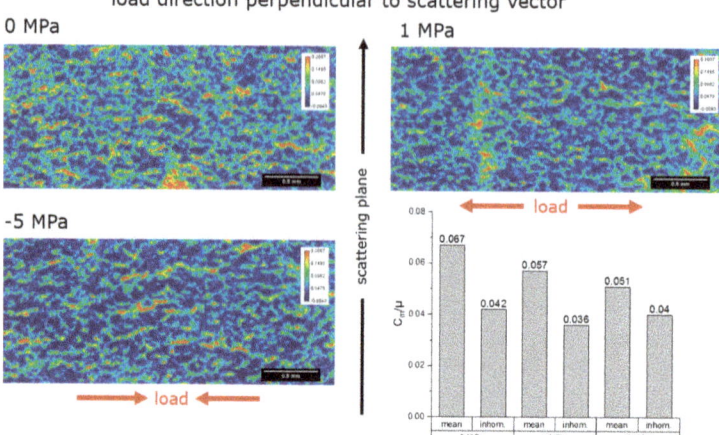

Figure 7. Visualization of the local values of the relative specific refraction value C_m/μ of the specimen as 2D color-coded images. The load direction of the specimen was perpendicular to the scattering plane of the analyzer crystal. Top left: Unloaded, top right: 1 MPa tensile loaded and bottom left: 5 MPa compression loaded. Bottom right: The integral values of the relative specific surface for all specimens are shown as bar graphs. The small bars represent the inhomogeneity of the C_m/μ values across the measured area of the specimen.

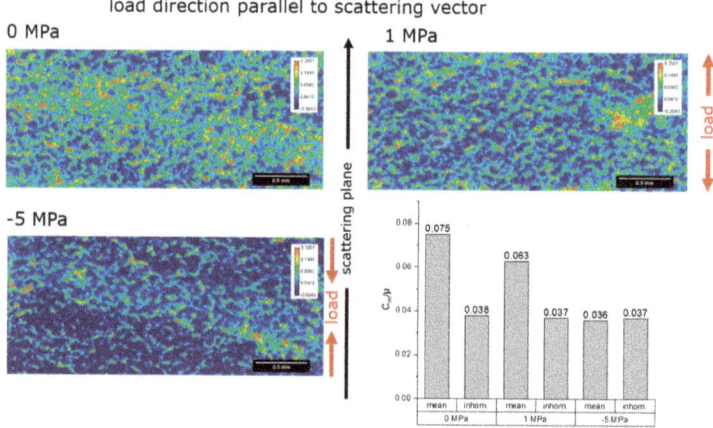

Figure 8. Visualization of the local values of the relative specific refraction value C_m/μ of the specimen as 2D color-coded images for the orientation of the load direction of the specimen parallel to the scattering plane of the analyzer crystal. Top left: Unloaded, top right: 1 MPa tensile loaded and bottom left: 5 MPa compression loaded. The load direction was parallel to the scattering vector. Top left: Unloaded, top right: 1 MPa tensile loaded and bottom left: 5 MPa compression loaded. Bottom right: The integral values of the relative specific surface for all specimens are shown as bar graphs. The small bars represent the inhomogeneity of the $\mu \cdot d$ values across the measured area of the specimen.

The 0 MPa specimen shows the highest specific surface in both orientations. The value of $C_m \cdot d$ was roughly the same for both orientations (no preferred orientation of the features causing refraction, namely grain boundaries and microcracks). The areas of high specific surface are localized.

The −5 MPa specimen shows the lowest specific surface. The value of $C_m \cdot d$ was higher for the orientation of the load axis perpendicular to the scattering plane (preferred orientation of features parallel to the load axis). Also, the local maxima are higher than for the other two specimens.

The 1 MPa specimen shows intermediate specific surface. The value of $C_m \cdot d$ was roughly the same for both orientations (no preferred orientation). The local maxima are similar to the 0 MPa specimen.

3.4. Quantification of Elastic Constants' Anisotropy

Since the arrangement of microcracks depends on the applied load during cooling (closure occurs in the case of compression, propagation in the case of tension), the degree of anisotropy of damage was quantified here through the measurement of the anisotropy of elastic constants (the microcrack induced anisotropy of properties has been predicted by Kachanov [32]). For this purpose, some entries of the stiffness tensor (C_{ij}) were determined on specimens cooled under different applied stress through ultrasonic transmission measurements in different directions. A uniaxial load is likely to induce a transverse isotropic symmetry, therefore, measurements were focused on the constants C_{11}, C_{22} and C_{33} (axis 3 is parallel to the applied load). For an isotropic material, the constants C_{11}, C_{22} and C_{33} should be equal, whereas in the case of transverse isotropy, C_{33} would be different to $C_{11} = C_{22}$. This anisotropy of the elastic constants was quantified (Figure 9b) through the index *AI*.

$$AI = \frac{C_{33}}{\left[\frac{C_{11}+C_{22}}{2}\right]} - 1 \qquad (3)$$

Figure 9 shows that:

- An unconstrained cooling yields a value of *AI* close to 0 (similar values of elastic constants in each direction), therefore to a rather isotropic microcrack arrangement;
- The application of a tensile stress during cooling leads to negative values of *AI* (C_{33} is smaller than C_{11} and C_{22}). This implies the generation of a network of microcracks that are preferentially oriented in the plane perpendicular to the direction of application of the load;
- The application of a compressive stress during cooling leads to positive values of *AI* (C_{33} is larger than C_{11} and C_{22}). This implies the generation of a network of microcracks that are preferentially oriented in the direction of the applied load.

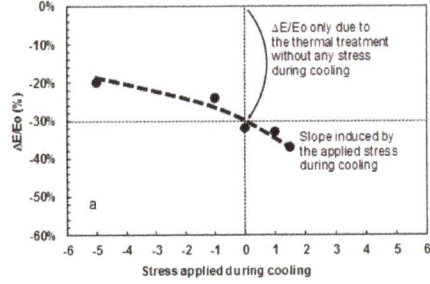
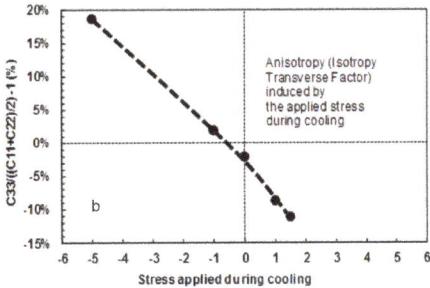

Figure 9. Room temperature elastic properties evolution due to thermal cycling (and cooling under applied stress, where axis 3 is the direction of the applied stress): (**a**) Decrease in Young's modulus measured along the direction of the applied load and (**b**) evolution of the anisotropy index.

In conclusion, these results suggest that the modulation of the deformation during the T → M transformation of zirconia under load could be related to the preferential direction of microcracks. This would establish a relationship between damage distribution (orientation) and the applied load during the T → M transformation that leads to the TRIP effect. In some other work, the orientation anisotropy of the microcrack arrangement has already been deduced (from lattice strain neutron diffraction measurements) and exemplarily observed by electron backscatter diffraction (EBSD) in porous Al_2TiO_5 by Bruno et al. [11]. Nevertheless, it was not possible in this case to clearly determine the preferential orientation of microcracks in SEM pictures, since their field of view was limited.

4. Discussion

It was expected that an external uniaxial stress should close microcracks oriented perpendicular to the load axis in the compression case, and open them in the tensile case. Correspondingly, we also expected that in the compression case, a possible rise of the opened microcracks in the direction parallel to the load axis would occur. Table 2, summarizing the X-ray refraction results, shows some expected results: In the parallel orientation of the load axis to the scattering plane, the decrease of refraction value in the −5 MPa specimen (with respect to the behavior of the 0 MPa specimen) corresponded to the decrease of microcrack density (or specific surface); in the perpendicular orientation of the load axis to the scattering plane, the (slight) decrease of refraction value for the +1 MPa specimen (with respect to the behavior of the 0 MPa specimen) corresponded to microcrack closure in the direction perpendicular to the load (Poisson's effect as microcrack lips come together and fall below the detection limit of the technique). However, some apparently surprising trends also appeared: In the parallel orientation of the load axis to the scattering plane, a slight decrease of the refraction value in the +1 MPa specimen occurred, which corresponded to a decrease of microcrack density (or specific surface); in the perpendicular orientation of the load axis to the scattering plane, the (slight) decrease of refraction value for the −5 MPa specimen occurred, which corresponded to microcrack closure in the direction perpendicular to the load. These two effects cannot be explained by Poisson's contraction (tension case) or expansion (compression case). A plausible explanation has been predicated in [15]: Damage in microcracked ceramics actually proceeds by propagation of existing microcracks, rather than by formation of new microcracks. This would imply that small microcracks can suddenly find themselves in the shielding zone of larger ones, thereby falling below the detection limit of X-ray refraction (~1 nm). Indeed, it has been shown in [15] that even under tension, some regions of a microcracked material undergoes local strain release, and when unloaded the detectable specific surface decreases even if the actual microcrack density (defined as $\rho = 1/V \cdot \sum_i a_i^3$, where V = investigated volume and a_i = radius of the i-th microcrack) increases, because of its cubic dependence on the crack size a. Furthermore, the 1 MPa specimen did possess a larger refraction value in the parallel orientation of the load axis to the scattering plane, i.e., a larger specific surface of microcracks oriented perpendicular to the applied load, but the amount of external tension is not enough to propagate existing microcracks to the same amount that a compressive stress of −5 MPa can do.

Table 2. Normalized global refraction value C_m/μ as a function of orientation of the load axis of the investigated specimens to the scattering plane of the analyzer crystal. Relative error bars lie around 1–2%. Note that cracks perpendicular to the load axis are visible if the scattering plane is parallel to the load axis and vice-versa.

Specimen	Load Axis Perpendicular to Scattering Plane	Load Axis Parallel to Scattering Plane
+1 MPa	0.057	0.063
0 MPa	0.067	0.075
−5 MPa	0.051	0.036

We also have to take into account that the quantitative analysis of the X-ray refraction maps of Figures 7 and 8 strongly depends on the segmentation procedure utilized to extract the refraction value. By applying different masks to the images, one can obtain slightly different results. In Figure 10 it is shown that different masks are obtained with different methods (see [33,34]). Those masks yield slightly different refraction values, as summarized in Table 3. Table 3 shows a similar trend to Table 2, with one important exception: Specimens 0 MPa and 1 MPa do not differ much. This analysis would rather support the hypothesis that indeed specimen 1 MPa did not undergo enough deformation to induce significant microcrack propagation (further damage to the initial condition), but this is subject to future work.

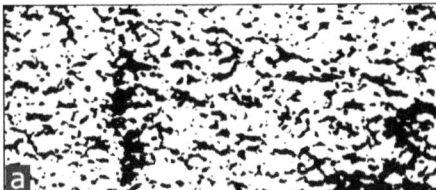 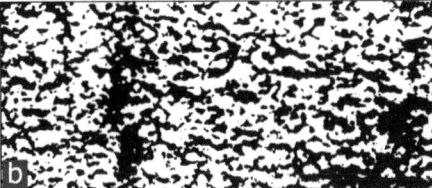

Figure 10. Segmentation of the refraction map for the 1 MPa specimen (Figure 7): (**a**) Mask with Otsu threshold; and (**b**) mask with Huang threshold.

Table 3. Normalized refraction value C_m/μ as a function of direction for the investigated specimens, calculated using two different segmentation masks (Otsu and Huang).

C_m/μ	Otsu		Huang	
Specimen	Load Axis Perpendicular to Scattering Plane	Load Axis Parallel to Scattering Plane	Load Axis Perpendicular to Scattering Plane	Load Axis Parallel to Scattering Plane
+1 MPa	0.048	0.060	0.042	0.047
0 MPa	0.061	0.062	0.046	0.057
−5 MPa	0.038	0.030	0.037	0.032

5. Conclusions

We have confirmed that synchrotron X-ray refraction is a useful technique to determine the evolution of damage, especially in brittle (microcracked) materials. While classically limited to light materials, we have expanded the use of X-ray refraction to a high-density material such as electro-fused refractory zirconia. We have shown that one can change the amount of microcracking in this material, a ZrO_2-SiO_2 composite, by means of an externally applied uniaxial stress during the cooling branch of a thermal cycle: A compressive load will close microcracks perpendicular to the applied load. This change therefore caused the anisotropy of the microcrack orientation. Upon application of a tensile load during cooling, microcrack propagation seemed to take place, whereby small cracks virtually closed (i.e., they fell below the detection limit of X-ray refraction techniques), however the X-ray refraction data can also be interpreted so that the investigated tensile load may not have induced enough damage to be detected. To clarify the issue, further investigations are needed.

Author Contributions: Conceptualization, M.H. and G.B.; Experimental Investigation, F.G, R.L. and B.R.M.; Data Curation, F.G., R.L. and B.R.M.; Writing—Original Draft Preparation, M.H. and G.B.; Writing—Review & Editing, F.G., R.L., B.R.M., G.B., G.A., T.C. and M.G.; Supervision, M.H., T.C., G.A. and G.B.; Project Administration, M.H. and T.C.; Funding Acquisition, M.H. and T.C.

Funding: This research was funded by the French National Research Agency grant number [ANR-12-RMNP-0007].

Acknowledgments: Authors are thankful to Saint-Gobain CREE for the supplying of materials. We thank HZB for the allocation of synchrotron radiation beamtime. We thank HZB colleagues for their support as well as Ralf Britzke and Thomas Wolk (BAM) for their assistance during beam time at BAMline.

Conflicts of Interest: The authors declare no conflict of interest.

References

1. Duvierre, G.; Boussant-Roux, Y. Fused Zirconia or Fused AZS: Which Is the Best Choice? In *59th Conference on Glass Problems: Ceramic Engineering and Science Proceedings*; American Ceramic Society: Hoboken, NJ, USA, 2019; pp. 65–80.
2. Garvie, R.C.; Hannink, R.H.; Pascoe, R.T. Ceramic Steel? *Nature* **1975**, *258*, 703–704. [CrossRef]
3. Patapy, C.; Gey, N.; Hazotte, A.; Humbert, M.; Chateigner, D.; Guinebretiere, R.; Huger, M.; Chotard, T. Mechanical behavior characterization of high zirconia fused-cast refractories at high temperature: Influence of the cooling stage on microstructural changes. *J. Eur. Ceram. Soc.* **2012**, *32*, 3929–3939. [CrossRef]
4. Kisi, E.H.; Howard, C.J. Crystal Structures of Zirconia Phases and their Inter-Relation. *Key Eng. Mater.* **1998**, *153–154*, 1–36. [CrossRef]
5. Badwal, S.P.; Bannister, M.J.; Hannink, R.H.J. *Science and Technology of Zirconia V*; Technomic Pub. Co.: Lancaster, PA, USA, 1993.
6. Bruno, G.; Efremov, A.M.; An, C.; Nickerson, S. Not All Microcracks are Born Equal: Thermal vs. Mechanical Microcracking in Porous Ceramics. *Ceram. Eng. Sci. Proc.* **2011**, *32*, 137–152. [CrossRef]
7. Thomas, H.A.J.; Stevens, R. Aluminium titanate—A literature review. Part 1. Microcracking phenomena. *Br. Ceram. Trans. J.* **1989**, *88*, 144–151.
8. Bruno, G.; Kachanov, M. Porous microcracked ceramics under compression: Micromechanical model of non-linear behavior. *J. Eur. Ceram. Soc.* **2013**, *33*, 2073–2085. [CrossRef]
9. Bruno, G.; Kachanov, M. Microstructure–Property Connections for Porous Ceramics: The Possibilities Offered by Micromechanics. *J. Am. Ceram. Soc.* **2016**, *99*, 3829–3852. [CrossRef]
10. Bruno, G.; Garlea, V.O.; Muth, J.; Efremov, A.M.; Watkins, T.R.; Shyam, A. Microstrain temperature evolution in b-eucryptite ceramics: Measurement and model. *Acta Mater.* **2012**, *60*, 4982–4996. [CrossRef]
11. Bruno, G.; Efremov, A.M.; Wheaton, B.R.; Webb, J.E. Microcrack orientation in porous aluminum titanate. *Acta Mater.* **2010**, *58*, 6649–6655. [CrossRef]
12. Gouraud, F. Influence des Transformations de phase de la Zircone sur le Comportement Thermomécanique de Réfractaires à Très Haute Teneur en Zircone. Ph.D. Thesis, University of Limoges, Limoges, France, 2016.
13. Patapy, C. Comportement Thermomécanique et Transformations de Phase de Matériaux Réfractaires électrofondus à Très Haute Teneur en Zircone. Ph.D. Thesis, University of Limoges, Limoges, France, 2010.
14. Kupsch, A.; Müller, B.R.; Lange, A.; Bruno, G. Microstructure characterisation of ceramics via 2D and 3D X-ray refraction techniques. *J. Eur. Ceram. Soc.* **2017**, *37*, 1879–1889. [CrossRef]
15. Müller, B.R.; Cooper, R.C.; Lange, A.; Kupsch, A.; Wheeler, M.; Hentschel, M.P.; Staude, A.; Pandey, A.; Shyam, A.; Bruno, G. Stress-induced microcrack density evolution in β-eucryptite ceramics: Experimental observations and possible route to strain hardening. *Acta Mater.* **2018**, *144*, 627–641. [CrossRef]
16. Müller, B.R.; Lange, A.; Harwardt, M.; Hentschel, M.P. Synchrotron-Based Micro-CT and Refraction-Enhanced Micro-CT for Non-Destructive Materials Characterisation. *Adv. Eng. Mater.* **2009**, *11*, 435–440. [CrossRef]
17. Fuller, E.R., Jr.; Granto, A.V.; Holder, J.; Naimon, E.R. 7. Ultrasonic Studies of the Properties of Solids. In *Methods in Experimental Physics*; Coleman, R.V., Ed.; Academic Press: Cambridge, MA, USA, 1974; Volume 11, pp. 371–441.
18. Hentschel, M.P.; Hosemann, R.; Lange, A.; Uther, B.; Bruckner, R. Small-Angle X-Ray Refraction in Metal Wires, Glass-Fibers and Hard Elastic Propylenes. *Acta Crystallogr. A* **1987**, *43*, 506–513. [CrossRef]
19. Müller, B.R.; Hentschel, M.P. Micro-diagnostics: X-ray and synchrotron techniques. In *Handbook of Technical Diagnostics—Fundamentals and Application to Structures and Systems*; Czichos, H., Ed.; Springer: Berlin/Heidelberg, Germany, 2013; pp. 287–300. [CrossRef]
20. Cooper, R.C.; Bruno, G.; Wheeler, M.R.; Pandey, A.; Watkins, T.R.; Shyam, A. Effect of microcracking on the uniaxial tensile response of β-eucryptite ceramics: Experiments and constitutive model. *Acta Mater.* **2017**, *135*, 361–371. [CrossRef]
21. Laquai, R.; Müller, B.R.; Kasperovich, G.; Haubrich, J.; Requena, G.; Bruno, G. X-ray refraction distinguishes unprocessed powder from empty pores in selective laser melting Ti-6Al-4V. *Mater. Res. Lett.* **2018**, *6*, 130–135. [CrossRef]
22. Görner, W.; Hentschel, M.P.; Müller, B.R.; Riesemeier, H.; Krumrey, M.; Ulm, G.; Diete, W.; Klein, U.; Frahm, R. BAMline: The first hard X-ray beamline at BESSY II. *Nucl. Instrum. Methods Phys. Res. Sect. A* **2001**, *467*, 703–706. [CrossRef]

23. Rack, A.; Zabler, S.; Müller, B.R.; Riesemeier, H.; Weidemann, G.; Lange, A.; Goebbels, J.; Hentschel, M.; Görner, W. High resolution synchrotron-based radiography and tomography using hard X-rays at the BAMline (BESSY II). *Nucl. Instrum. Methods Phys. Res. Sect. A* **2008**, *586*, 327–344. [CrossRef]
24. Schneider, C.A.; Rasband, W.S.; Eliceiri, K.W. NIH Image to ImageJ: 25 years of image analysis. *Nat. Methods* **2012**, *9*, 671. [CrossRef] [PubMed]
25. Nellesen, J.; Laquai, R.; Müller, B.R.; Kupsch, A.; Hentschel, M.P.; Anar, N.B.; Soppa, E.; Tillmann, W.; Bruno, G. In situ analysis of damage evolution in an Al/Al2O3 MMC under tensile load by synchrotron X-ray refraction imaging. *J. Mater. Sci.* **2018**, *53*. [CrossRef]
26. Cabeza, S.; Müller, B.R.; Pereyra, R.; Fernandez, R.; Gonzalez-Doncel, G.; Bruno, G. Evidence of damage evolution during creep of Al–Mg alloy using synchrotron X-ray refraction. *J. Appl. Crystallogr.* **2018**, *51*, 420–427. [CrossRef]
27. Patapy, C.; Huger, M.; Guinebretière, R.; Gey, N.; Humbert, M.; Hazotte, A.; Chotard, T. Solidification structure in pure zirconia liquid molten phase. *J. Eur. Ceram. Soc.* **2013**, *33*, 259–268. [CrossRef]
28. Krogstad, J.A.; Gao, Y.; Bai, J.; Wang, J.; Lipkin, D.M.; Levi, C.G. In Situ Diffraction Study of the High-Temperature Decomposition of t′-Zirconia. *J. Am. Ceram. Soc.* **2015**, *98*, 247–254. [CrossRef]
29. Mitter, W. *Umwandlungsplastizität und ihre Berücksichtigung bei der Berechnung von Eigenspannungen*; Gebrüder Borntraeger: Berlin/Stuttgart, Germany, 1987.
30. Greenwood, G.W.; Johnson, R.H.; Rotherham, L. The deformation of metals under small stresses during phase transformations. *Proc. R. Soc. Lond. Ser. A Math. Phys. Sci.* **1965**, *283*, 403–422. [CrossRef]
31. Magee, C.L.; Paxton, H.W. Transformation Kinetics, Microplasticity and Aging of Martensite in Fe-31Ni. Ph.D. Thesis, Carnegie Institute of Technology Pittsburgh, Pittsburgh, PA, USA, 1966.
32. Kachanov, M. Elastic Solids with Many Cracks and Related Problems. In *Advances in Applied Mechanics*; Hutchinson, J.W., Wu, T.Y., Eds.; Elsevier: Amsterdam, The Netherlands, 1993; Volume 30, pp. 259–445.
33. Otsu, N. A Threshold Selection Method from Gray-Level Histograms. *IEEE Trans. Syst. Man Cybern.* **1979**, *9*, 62–66. [CrossRef]
34. Huang, L.-K.; Wang, M.-J.J. Image thresholding by minimizing the measures of fuzziness. *Pattern Recognit.* **1995**, *28*, 41–51. [CrossRef]

© 2019 by the authors. Licensee MDPI, Basel, Switzerland. This article is an open access article distributed under the terms and conditions of the Creative Commons Attribution (CC BY) license (http://creativecommons.org/licenses/by/4.0/).

Article

Instrumented Indentation of Super-Insulating Silica Compacts

Belynda Benane [1,2,3], **Sylvain Meille** [1,2,*], **Geneviève Foray** [1,2], **Bernard Yrieix** [2,3] **and Christian Olagnon** [1,2]

1. Univ Lyon, INSA-Lyon, MATEIS, UMR CNRS 5510—7 avenue Jean Capelle, F-69621 Villeurbanne, France; belynda-b.benane@edf.fr (B.B.); genevieve.foray@insa-lyon.fr (G.F.); christian.olagnon@insa-lyon.fr (C.O.)
2. Univ Lyon, INSA-Lyon, CNRS, UCBL, MATEB, 7 avenue Jean Capelle, F-69621 Villeurbanne, France; bernard.yrieix@edf.fr
3. EDF R&D, Les Renardières, F-77250 Moret sur Loing, France
* Correspondence: sylvain.meille@insa-lyon.fr; Tel.: +33-472438064

Received: 10 February 2019; Accepted: 7 March 2019; Published: 12 March 2019

Abstract: Highly porous silica compacts for superinsulation were characterized by instrumented indentation. Samples showed a multi-scale stacking of silica particles with a total porous fraction of 90 vol %. The two main sources of silica available for the superinsulation market were considered: fumed silica and precipitated silica. The compacts processed with these two silica displayed different mechanical properties at a similar porosity fraction, thus leading to different usage properties, as the superinsulation market requires sufficient mechanical properties at the lowest density. The measurement of Young's modulus and hardness was possible with spherical indentation, which is an efficient method for characterizing highly porous structures. Comparison of the mechanical parameters measured on silica compacts and silica aerogels available from the literature was made. Differences in mechanical properties between fumed and precipitated compacts were explained by structural organization.

Keywords: silica; super-insulating materials; instrumented indentation; porosity

1. Introduction

Reducing energy consumption in buildings is a critical issue, as it represents more than 40% of the total energy consumption in the so-called developed countries [1,2]. To drastically reduce the energy needed for heating and cooling purposes in buildings, new classes of super-insulating materials are needed, allowing either to ensure a higher insulation capacity as compared to standard insulation materials or reduce the current thicknesses of insulating materials at a given insulation capacity. The most promising solution is currently vacuum insulation panels (VIP), which are made of a core of slightly compacted silica nanopowders (with typically 90% of porosity) wrapped into a sealed membrane under vacuum. Thermal conductivities as low as 2 to 5 mW/(m·K) can be achieved for these structures, which is eight times lower than conventional insulation materials such as expanded polystyrene or mineral wool [3]. However, a major limitation to the commercial development of VIP is their high price. To reduce it, the preferential solution would be to replace the fumed silica (FS) that is currently used as the core material with precipitated silica (PS), which is largely used in many industrial applications (tires, pharmaceutics, etc.). The former are synthesized at high temperature, the latter in aqueous solution at a lower price and in larger quantities. The choice of the nature of the silica nanopowder has a major influence on the mechanical characteristics and on the thermal conductivity of VIP [4–6]. At a given compaction pressure, PS compacts are denser and therefore have a higher thermal conductivity as compared with FS compacts [7], thus limiting the development of VIP with a core made with PS. Therefore, it is of critical importance to understand the origins of such differences in the mechanical behavior between compacts of PS and FS powders. This could open new avenues to either develop more competitive solutions in superinsulation or improve existing solutions.

The structural characteristics of PS silica nanopowders have been largely studied using small angles X-ray scattering (SAXS) [8,9] and transmission electronic microscopy (TEM) [10]. It was shown that these powders have a pronounced multi-scale structure that is made of elementary silica nanoparticles (5 to 20 nm in diameter) organized at larger scales in aggregates (in the 100 to 200-nm size range) and agglomerates (400 to 500-nm size range). The nanometer pore size combined with the very open structure of the silica and the presence of free dangling arms limit the thermal conduction through the gas and the solid phases. FS powder and compacts were recently studied using SAXS, TEM, and mercury instruction porosimetry [11]. The multi-scale organization already noted for PS was confirmed. Moreover, the non-spherical shape of aggregates of FS particles (necklace-shaped and disk-shaped) at the submicron-length scale was highlighted.

Whereas numerous publications deal with the mechanical properties of highly porous ceramics and glasses in the range of 50% to 75% of porosity [12–19], the literature offers a limited number of papers dealing with the mechanical properties of highly porous mineral materials. Most studies deal with the mechanical properties of highly porous silica in the form of aerogels that are used as super-insulating material at atmospheric pressure, with a range of density from 0.08 to 0.35 g/cm^3, i.e., 4% to 15% of solid fraction, which is similar to or lower than that for silica compacts (typically 10%) [20,21]. Silica aerogels have been characterized in terms of their elastic modulus, compression strength, and fracture toughness for different apparent densities [22,23]. The preparation of samples for standard mechanical testing was noted as delicate in such materials with very low mechanical properties [21,23]. Instrumented indentation has also been used to characterize the properties of silica aerogels, namely elastic modulus and hardness. Typical values of elastic modulus and hardness depend on the solid content, E, varying from 1 to 100 MPa, and H varying from 0.1 to 10 MPa for solid fractions ranging from 5 to 20 vol % [21,23]. Ultra-low density aerogels (apparent density below 0.04 g/cm^3, i.e., 2% solid fraction) have also been characterized [21,24]. Their mechanical behavior in spherical indentation tends to be mainly elastic with a large deformation capacity [25]. In addition to the influence of the structure on properties, an influence of the relative humidity (RH) on the mechanical behavior of highly porous silica aerogels has also been noted, with an increase of the time-dependent contribution at high RH [26]. Silica powders in the form of dense compacts for pharmaceuticals have been studied, but at relative densities above 30%, which is much larger than the density that is needed for the superinsulation market [27]. To our knowledge, no work has been published on the mechanical properties of silica compacts with densities in the range of superinsulation materials.

This work focuses on the characterization of the mechanical properties of compacts of nanostructured silica powders to be used as the core material of VIP. Two nature of silica were considered: fumed silica (FS) and precipitated silica (PS). Due to the high volume fraction of the porosity in such structures, their mechanical properties are very low, and the preparation of samples with well-defined geometries for standard tests to characterize mechanical properties (flexion, compression) is delicate. Uniaxial compression strength is also dependent on the aspect ratio of samples, as the friction between platens and sample surface can influence the measurement of strength. Instrumented indentation offers a credible alternative due to its relative easy setup, quasi-nondestructive character, and the ability to run several tests on a single sample.

Indentation tests were carried out with sharp and spherical tips at a large penetration depth to test a large volume of material, allowing averaging the contributions of both solid and porous phases in the material [28–32]. Test were carried out on both FS and PS silica compacts at the same total porosity fraction to help understand the differences in the properties noted in VIP formulations between the two sources of silica. All of the tests were performed at ambient conditions to limit the influence of water absorption on the properties of materials.

2. Materials and Methods

Two commercial nanostructured silica powders were characterized—a fumed silica (Konasil 200, OCI, Seoul, Korea) and a precipitated silica (T43, Solvay, Collonges-au-Mont-d'Or, France)—both with purity above 99.5%. Cylindrical pellets of FS and of PS powders were fabricated by oedometric compression on a Zwick testing machine (BZ1-MM1195, Zwick-Roell, Ulm, Germany) equipped with a steel die (inner diameter of 20 mm) and a brass piston. The same weight of FS or PS powder (200 mg) was introduced in the die before compaction. The surface roughness of the sample was controlled by a disk of mirror-polished sapphire (20-mm diameter and 2-mm thick) placed inside the die below the silica compact. The crosshead speed during loading and unloading were 10 mm·mn^{-1} and 5 mm·mn^{-1} respectively; a one-hour stress relaxation stage was made after the loading phase, as it promotes the flattening of the sample's surface. A typical roughness Ra of 1.5 µm was measured on the tested sample's surface using a Hirox KH-7700 digital microscope (Hirox-Europe, Limonest, France). A similar volume fraction of pores was targeted for both FS and PS silica samples for the sake of comparison in mechanical properties (10% of solid volume fraction targeted, i.e., an apparent density of 210–220 g/m^3). To reach the targeted apparent densities after compaction, maximum applied pressures were 0.6 MPa and 0.15 MPa for FS and PS respectively, illustrating the easier densification of PS powder as compared with FS powders. The final sample diameter and height were 20 mm and 3 mm, respectively. All of the compaction tests were performed at 20 °C and between 30–45% RH (measured by a thermohygrometer, Testo 625, Testo, Lenzkirch, Germany) to limit the influence of environmental conditions. Table 1 gathers the main physical parameters measured on FS and PS silica powders and on FS and PS pellets.

Table 1. Main physical characteristics of fumed silica (FS) and precipitated silica (PS) compacts (average values and standard deviations when available) as determined in this work. Water uptake is measured on silica powders; after drying at 140 °C for 1 h, skeletal density is measured with a helium pycnometer (Accu-Pyc, Micrometritics, Norcross, GA, USA), and specific surface by nitrogen absorption (Belsorp-max, BEL, Germany). Numbers in italic stand for the standard deviation.

Silica	Solid Fraction	Specific Surface	Skeletal Density	Apparent Density	Water Uptake (wt %)
	%	g/m^2	kg/m^3	kg/m^3	24 °C 45% RH
FS	9.9	187 *15*	2200	218 *2*	0.5 *0.1*
PS	9.4	207 *20*	2100	198 *2*	6.0 *0.4*

Instrumented indentation tests were performed on a nanoindenter G200 (Agilent, Santa Clara, CA, USA), equipped with a continuous stiffness measurement (CSM) module. Synthetic sapphire spherical indenter tips were used with diameters of 600 µm and 2000 µm, respectively, where are hereafter labeled D600 and D2000. The actual sphere radius of curvature was measured with an AFM (Dimension 3100, Veeco, Plainview, NY, USA) to calculate the exact tip area functions. Some tests were also performed with a diamond Berkovich tip. The calibration of the tip area function was made using a standard fused silica sample. Indentations were carried out under a constant strain rate of 0.1 s^{-1} to a maximum depth of 5 µm for the Berkovich tip and 20 µm for the spherical tips. The load was also maintained at its maximal value for 10 s before unloading the material. Temperature and relative humidity conditions were set to 20 °C with RH varying from 30% to 45%.

The determination of the contact point was made using the method described by Moseson et al. [33] and successfully applied to porous ceramics [29,34]. E was determined using both CSM and Oliver and Pharr method [35]. The Poisson's ratio, ν, was set to 0.2, which was a value obtained by Sanahuja et al. for highly porous materials, independently of the Poisson's ratio of the solid matrix [19]. Three different samples for each silica were tested with a minimum of six indents per sample. To avoid proximity effects, the minimal center-to-center distance between two neighbor indents was set at four times the radius of contact for spheres and four times the side of residual imprints for the Berkovich tip. The observation of the residual imprints was found to be difficult due to the large radius of curvature of the spheres used and the uniform white color of the samples. Gold coating prior to SEM or optical microscopy observation was needed.

A spherical indentation test enabled the determination of contact stress–strain behavior [21,28,36] by plotting the mean pressure of hardness p_m, which was defined as: $p_m = \frac{P}{\pi a^2}$ as a function of the indentation strain a/R, where a is the radius of contact and R is the sphere radius. As a spherical tip avoids stress singularities, an elastic contact is likely to occur at low loads. The theory of elastic contact, as suggested by Hertz, prescribes a linear relation between p_m, the indentation stress, and a/R, the indentation strain: $p_m = \frac{4E_r}{3\pi} \cdot \frac{a}{R}$ where P is the indentation load, and E_r is the reduced modulus defined by: $\frac{1}{E_r} = \frac{1-\nu_m^2}{E_m} + \frac{1-\nu_i^2}{E_i}$, where E and ν are the Young's modulus and Poisson's ratio of the tested material and of the indenter. This usually corresponds to the initial stage of the stress–strain curves in real experiments [36] and permits determining the Young's modulus E of the material (hereafter referred as the Hertz method for the determination of E).

Another parameter of interest is the ratio of the area under the unloading curve to the area under the loading curve, defined as the elastic-to-total work ratio W_e/W_{tot}. This ratio ranges from zero (totally irreversible behavior) to one (totally elastic behavior).

3. Results

The load–displacement curves measured on FS and PS samples using Berkovich, D600, and D2000 spheres are shown in Figure 1. The curves measured with Berkovich tips (Figure 1a) show a large scattering for both of the tested materials. Tests with D600 and D2000 (Figure 1b,c) spheres led to a much better repeatability of the load–displacement curves. Then, a clear distinction can be made between FS and PS compacts with a higher hardness for FS pellets as compared with PS, which was emphasized in the tests with D2000 spheres. The comparison of load–displacement curves for the three different tips shows a large influence of tip characteristics on the residual penetration depth after unloading (Figure 1). The Berkovich test led to larger irreversible fractions, which were illustrated by a large residual penetration depth compared with the maximum penetration depth. The irreversible energy fractions were lower in tests with D600 spheres, and even lower in tests with D2000 spheres. The latter show a strong elastic behavior, as illustrated by the evident elastic recovery during unloading with a limited residual penetration depth.

As the spherical indentation tests did not lead to a unique hardness value, contact stress–strain curves were plotted from the D2000 tests and are shown in Figure 3. They show a first linear relationship between hardness and a/R, allowing the determination of Young's modulus by the Hertz method, followed by a non-linear increasing part. Then, the average pressure at the beginning of non-linearity could be extracted. The parameters for FS and PS pellets determined from instrumented indentation tests, including Young's modulus, hardness, the elastic-to-total work ratio, are presented in Tables 2 and 3.

The Young's modulus determined with the CSM, Oliver and Pharr, and Hertz methods led to similar values for spherical indentation tests for both FS and PS pellets (Table 2). E values were slightly dependent on the size of spheres; they were higher with D2000 as compared with D600 for the FS samples, and higher with D600 for the PS samples. The strong influence of using a sharp tip as compared with a spherical one was noted, with much larger values of E for FS and PS compacts using a Berkovich tip. Comparing the two materials, FS compacts were approximately twice as stiff as PS from the D2000 tests, and only 25% higher from tests using D600 spheres. Berkovich indentation led to similar values of Young's modulus for the two silica samples, showing even slightly higher values for PS compacts.

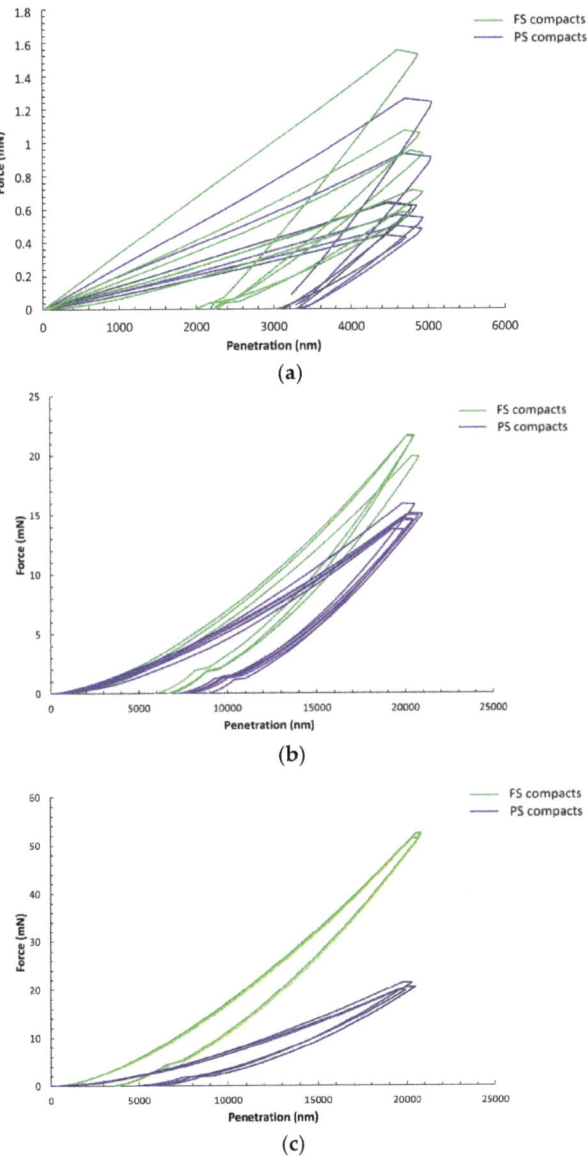

Figure 1. Load–deflection curves in spherical indentation on fumed silica (FS, in green) and precipitated silica (PS, in blue) compacts. (**a**) Berkovich tip; (**b**) 600-μm diameter sphere; and (**c**) 2000-μm diameter sphere.

The Young's modulus versus penetration depth (CSM mode) in the D2000 tests is shown in Figure 2. The average modulus was extracted in the plateau region, which was above 5 μm of penetration depth, thus avoiding the influence of the surface roughness of the samples. The higher rigidity of the FS compacts as compared with PS ones was clearly demonstrated.

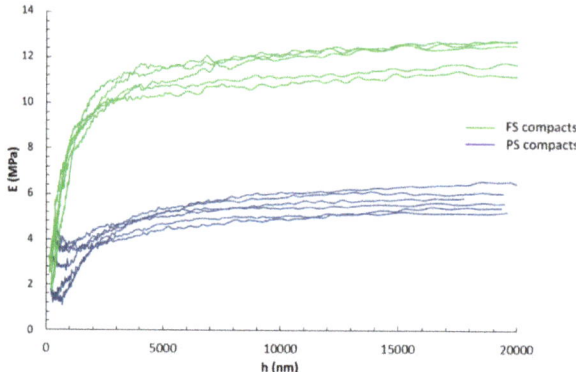

Figure 2. Young's modulus as measured by CSM versus penetration depth with a 2000-μm diameter sphere for FS (in green) and PS (in blue) compacts.

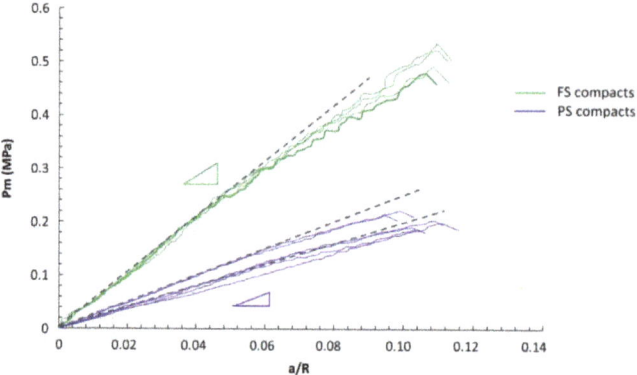

Figure 3. Contact stress–strain curves on FS (in green) and PS compacts (in blue) measured with a 2000-μm diameter sphere. Triangles illustrate the difference in slope between FS and PS compacts.

Table 2. Average values and standard deviations of elastic modulus of FS and PS compacts determined from indentation tests with different tips and using different calculation methods. Poisson's ratio is considered equal to 0.2. CSM: continuous stiffness measurement, OP: Oliver–Pharr. Numbers in italic stand for the standard deviation.

E (MPa)	Sphere						Berkovich
	D2000			D600			
	CSM	OP	Hertz	CSM	OP	Hertz	CSM
FS	11.8 $_{0.6}$	12.6 $_{0.6}$	12.3 $_{1.0}$	10.0 $_{0.6}$	10.7 $_{0.6}$	10.6 $_{1.0}$	21 $_{10}$
PS	5.5 $_{0.4}$	6.1 $_{0.4}$	5.7 $_{1.0}$	8.2 $_{0.3}$	9.1 $_{0.3}$	7.1 $_{1.0}$	24 $_{10}$

In terms of hardness, tests with Berkovich tips led to much higher values as compared with spherical tests. As spherical tests led to non-unique hardness values, the hardness indicated in Table 3 for the spherical tests corresponds to the pressure at the end of the elastic domain in the contact stress–strain curves (see Figure 3). Tests with D600 and D2000 spheres led to slightly different values of pressure for PS compacts, whereas for FS compacts, the values were identical. The comparison between FS and PS confirmed the large difference in properties already noted for the Young's modulus, with a higher hardness for FS compacts, which was particularly evidenced for tests with D2000 spheres

(FS pellets more than twice as hard as PS ones), and slightly less pronounced for tests with D600 spheres and Berkovich tips (FS pellets 40% harder than PS ones).

Table 3. Average values and standard deviations of elastic modulus, hardness, and elastic-to-total work ratio determined from indentation tests with spherical and Berkovich tips, using the CSM method. The values of hardness for the D2000 and D600 spherical tests correspond to the end of the elastic domain in the contact stress–strain curves. Numbers in italic stand for the standard deviation.

Property	Sample	Sphere		Berkovich
		D2000	D600	
E (MPa)	FS	11.8 $_{0.6}$	10.0 $_{0.6}$	21 $_{10}$
	PS	5.5 $_{0.4}$	8.2 $_{0.3}$	24 $_{10}$
H (MPa)	FS	0.35 $_{0.01}$	0.35 $_{0.02}$	5.2 $_{2.4}$
	PS	0.15 $_{0.01}$	0.25 $_{0.02}$	3.8 $_{1.8}$
W_e/W_{tot}	FS	0.77 $_{0.02}$	0.65 $_{0.03}$	0.45 $_{0.08}$
	PS	0.65 $_{0.02}$	0.55 $_{0.03}$	0.33 $_{0.06}$

The elastic-to-total work ratio was largely dependent on the tip geometry, with the largest values for the D2000 sphere, and lowest values for the Berkovich tip. This is clearly illustrated in the load–displacement curves (Figure 1) by the ratio of areas under the unloading and under the loading curves, respectively. Regarding the two materials tested, FS compacts showed a higher elastic-to-total work ratio as compared with PS compacts, whatever the testing conditions.

Figure 4 illustrates the residual imprints observed with SEM. The high level of porosity gives a heterogeneous surface aspect with some charging effect, making the observation delicate, especially for tests with D2000 with a strong elastic recovery at unloading and a small residual depth. No pile-up was noted around the residual imprints, whatever the tip geometry. No macrocracks were found at the corner of Berkovich residual imprints (Figure 4d,e). Spherical imprints did not show macrocracks; however, ring cracks could be noted above a threshold load (approximately 10 mN for tests with D600) on FS samples (Figure 4c), but not on PS samples.

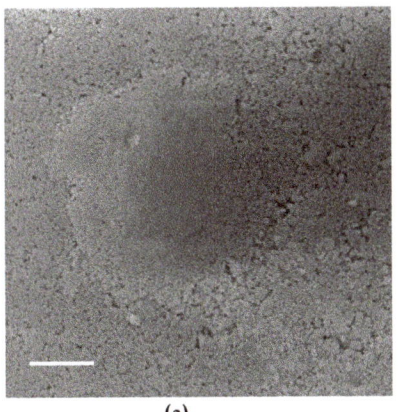
(a)

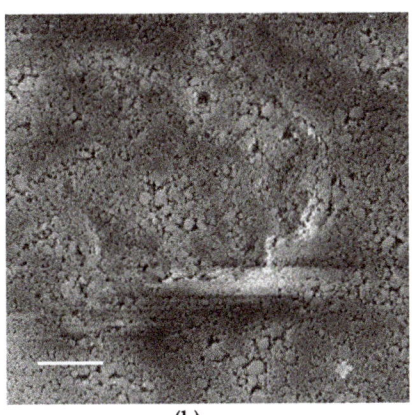
(b)

Figure 4. *Cont.*

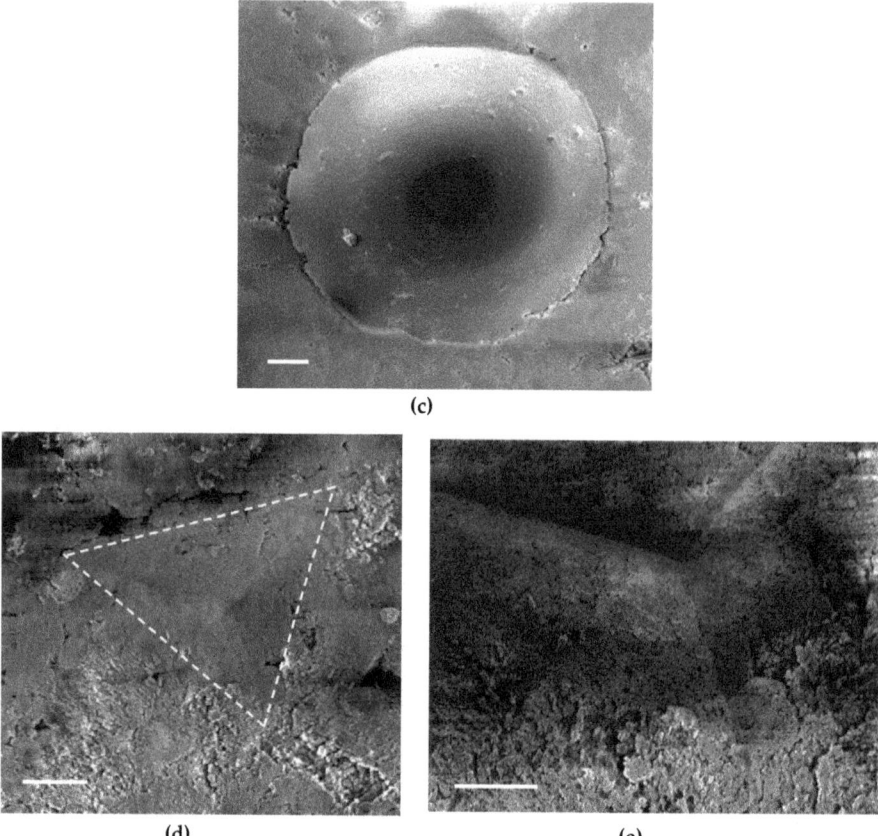

Figure 4. Images of residual indents: (**a**) D600 sphere on an FS compact, load 6 mN, (**b**) D600 sphere on a PS compact, load 6 mN, (**c**) D2000 sphere on an FS compact, load 50 mN, (**d**,**e**) Berkovich on an FS compact, load 1 mN (dotted lines in d added as guidelines). Scale bar: (**a**) 20 µm, (**b**) 20 µm, (**c**) 20 µm, (**d**) 20 µm, and (**e**) 10 µm.

4. Discussion

4.1. Porous Silica Compacts, A Peculiar Material

The mechanical testing of highly porous silica compacts is a tough task due to the difficulty of manipulating samples with a very low stiffness and hardness. Instrumented indentation offers an advantage over standard mechanical tests (compression, bending) due to the possibility to test samples with non-standard geometry, requiring only two parallel and flat surfaces. However, the impossibility of mechanically polishing the surface without degradation (cracks, densification) requires the development of a specific preparation protocol. In this study, samples were processed by oedometric compression with a one-hour stress relaxation stage between loading and unloading to decrease the surface roughness of the sample. Preliminary tests (not shown) have shown that mean values of Young's modulus and hardness are not affected by the relaxation stage, but that the experimental scatter was largely reduced after relaxation. The relaxation stage did not generate any dense "skin" layer at the sample's surface, as illustrated in Figure 2 by the constant value of Young's modulus versus penetration depth once the contact has been established. The white mat surface aspect of the samples, as well as their high porosity level, made the observation of residual imprints difficult with surface charging even after metal coating (Figure 4).

No piling up was noted neither in the spherical nor in the Berkovich tests, as already encountered on highly porous ceramics [28], which was related to the densification capacity of the material due to its high porous fraction. The densification of silica pellets in oedometric compaction at increasing pressures (0.15 to 1.2 MPa) has already been noted for FS samples and PS samples [11,37]. Some circumferential cracks could also be noted on FS compacts on the periphery of residual imprints for spherical tests for the highest loads used in this work (Figure 4). No macrocrack propagation from the corners of Berkovich residual imprints was noted, which can also be related to the densification of the material under the indent, reducing the residual tensile stress field at the origin of crack propagation after unloading [37]. This lower sensitivity to crack propagation from indents in highly porous ceramics has already been noted [38,39].

4.2. Influence of the Tip Geometry

The mechanical parameters extracted from the instrumented indentation tests, hardness, and Young's modulus varied largely with the tip characteristics, which included the geometry (sharp or spherical) and size for spherical tips.

Tests with Berkovich tips led to much higher E and H values than with spherical tips (Tables 2 and 3) as well as to a lower elastic contribution, as testified by the lower elastic-to-total work ratio (Figure 1, Table 3). Irreversible phenomena below the indenter were favored by the high stress concentration generated by Berkovich tips, as compared with spherical tips.

Comparing the spherical tests with D600 and D2000 spheres, a small difference was noted in the pressure at the end of the elastic domain of the contact stress–strain curves for PS compacts only, with average values of 0.15 MPa and 0.25 MPa for D2000 and D600, respectively (Figure 3, Table 3). This may be related to a different stress field under the indenter for a given a/R value between the two sphere sizes for a damageable material, as already noted in the literature on porous ceramics and mineralized bone tissue [29,39]. Nevertheless, it should be noted that spherical tests are much less sensitive to damage than the Berkovich test, which is due to the smoother stress fields as compared with sharp indenters [31,39].

The samples tested in this work were highly porous (90% of pores) and displayed a granular structure, which was made of a stacking of nanosized silica particles. The presence of an elastic domain for such material may be questioned. From the results of this work, it appears that a Young's modulus can be determined from spherical indentation tests. Indentation tests with a given sphere diameter gave similar values for E with different methods (Table 2), which included the Hertz and Oliver Pharr methods based on loading and unloading curves recorded at a static strain rate, and the CSM method on a high-frequency measurement, thus validating the existence of an elastic domain for both FS and PS compacts in ambient environmental conditions. A small difference was noted when comparing the Young's modulus extracted from D600 and D2000 tests, with a lower value for the FS modulus and a higher value for PS with D600 as compared with D2000. It seemed that tests with the D600 sphere showed a more pronounced tendency to involve irreversible mechanisms as compared with tests with the D2000 sphere, as testified by the lower elastic-to-total work ratio (Table 3). As a consequence, the Young's modulus calculated from the D600 tests may be influenced by damage, as noted to a larger extent on tests with Berkovich tips. Finally, D2000 spheres appear to be the preferred geometry to characterize compacts of silica nanopowders.

4.3. Comparison between Fumed and Precipitated Silica

From the experimental data collected in this work, a large difference in terms of mechanical properties is evidenced between compacts of FS and PS silica with a similar total fraction of pores. The comparison of samples is made at a similar apparent density, as density is a first-order parameter on the properties of porous materials [12,40]. FS compacts offered higher rigidity and hardness as compared with PS compacts, with a Young's modulus of 12 MPa and an average pressure at the end of elastic domain of 0.35 MPa versus 6 MPa and 0.15 MPa respectively (data from D2000 tests). This is consistent with previous work on the formulation of the VIP core [7] that showed the superior mechanical behavior of FS materials over PS in VIP core formulations. It is also consistent with the observation during handling that FS pellets are stronger than PS pellets, which tend to collapse under finger pressure.

The larger rigidity and pressure at the end of the elastic domain of FS compacts (more than twice the value of PS compacts) may be explained by the structural organization of the stacking of silica powders. The multi-scale organization of similar compacts was characterized by transmission electron microscopy and by small angle X-ray scattering (SAXS) experiments [11]. The organization at the submicron-length scale appeared as clearly different between FS and PS compacts. Silica agglomerates of FS silica were shown to have a high aspect ratio, which was linked with a stronger entanglement capacity, and finally with a stronger material as a given total porosity fraction [12,41]. When comparing the hardness and rigidity of silica pellets, it has to be noted that FS and PS samples were compacted at two different pressures (0.6 MPa for FS and 0.15 MPa for PS) to prepare samples with a similar apparent density. The different compaction behavior between FS and PS powders can also be related to the stronger entanglement between agglomerates of silica particles in FS samples as compared with PS ones. Compaction pressures are of the same order of magnitude as the mean pressures at the end of elastic domain measured with spherical indentation (0.3 MPa for FS and 0.15 MPa for PS). However, it needs to be noted that the stress field in oedometric compaction and spherical indentation tests are different, and a direct comparison of the pressure level is not possible.

When tested with Berkovich tips, smaller differences in E and H values were noted between FS and PS samples as compared with spherical tests (Table 3). This has to be related to the large irreversible phenomenon contribution, and mainly to a strong densification under the indenter, as testified by the large values of E and H with sharp tip as compared with spheres. It seems that once the stacking of silica agglomerates locally collapses and subsequently densifies, the differences between FS and PS samples are greatly reduced. It also appears that E measured using a Berkovich tip was not representative of the virgin material, showing the strong influence of the densification phenomena or damage. This tendency was already noted for bone tissue, which is a damage-sensitive material [39].

4.4. Comparison with Other Porous Silica Samples and Silica Aerogels

As mentioned in the introduction, no previous indentation tests on compacts of silica nanopowders were found in the literature. Instrumented indentation tests were carried out on silica aerogels [20,21,25] and foamed silica [42], which had apparent densities close to those of the silica compacts tested in this work.

The instrumented indentation of silica macroporous scaffolds with a solid fraction of 18 vol % and 24 vol % have shown that the major mechanism occurring during indentation is densification by local fracture of the solid walls between pores [42], with a very limited elastic contribution in tests with both Berkovich and spherical tips. Silica aerogels with a similar solid fraction as the silica compacts tested in this study (apparent density of 340 g/m^3 i.e., a solid fraction 15 vol % [21]) were also tested using Berkovich and spherical tips at room temperature. Young's modulus and hardness measured with both tip geometries led to similar values i.e., 110 MPa for E and 10 to 11 MPa for H. These E and H values of silica aerogels were larger than those noted for the silica compacts tested in this study (Table 3). The mechanism involved during the indentation of silica aerogels is the bending of nanoligaments of silica with no major signs of densification, but with the presence of cracks from the corner of the residual imprint in Berkovich tests, and of numerous ring cracks in spherical tests [21]. The 3D distribution of the solid phase in aerogel and silica compacts is different, with a 3D network of small size (3 to 6-nm) silica ligaments for aerogels [43,44] versus a multi-scale stacking of aggregates and agglomerates for silica compacts [11]. The former give a pore size distribution centered on 10 nm, while the latter give a multi-scale pore distribution with pores from several nanometers to tens of micrometers [11]. The stronger resistance to the densification of silica aerogels as compared with silica compacts may be related to the different organization of the solid phase and to the presence of Si–C bonds, leading to a better strength and a higher amount of stored elastic energy, and therefore to a more brittle behavior.

4.5. Final Remarks, Outlook

A fine characterization of the mechanical properties of VIP core is of critical importance, as these materials need to show sufficient mechanical properties to enable the fabrication of panels at the lowest possible density in order to limit the thermal conductivity. It appears from this study that indentation testing with a large sphere offers the best choice to extract elastic properties and estimate the average pressure at the end of elastic domain with a large tested volume.

However, the characterization of the damaged volume below the residual indent needs to be improved, possibly through using FIB slice and viewing the observations. In particular, the quantification of densification would bring valuable information for identifying the constitutive law of the material, as already demonstrated for densifying glasses or porous ceramics [29,45,46]. Modeling the mechanical properties of a multi-scale stacking of silica particles [43] also appears as critical to better understand the structure–properties relationship of silica compacts, similarly to that which has been done for silica aerogels [44].

We have chosen in this work to maintain limited and similar environmental conditions by working in an ambient environment under controlled temperature and with only small variations in relative humidity. Large differences exist in the hydrophilicity of silica nanopowders, which are related to their processes and possible surface treatments (see Table 1 [5,47]). Thus, an influence of the water absorption on the mechanical properties of compacts is expected. It could be useful to test the compacts at 140 °C to eliminate the physisorbed water [48,49], and therefore get closer to the conditions in the core of a VIP. At the same time, it is also of importance to characterize the aging of silica compacts versus relative humidity, as the pressure inside the VIP is known to increase during the service life of the insulating materials [49].

5. Conclusions

In this work, we have demonstrated the relevance of spherical indentation to characterize the mechanical properties of highly porous silica compacts with a porosity fraction of 90% and a very low rigidity, with a Young's modulus ranging between 5–10 MPa. This experimental technique allows discriminating the properties of two types of silica powder (fumed and precipitated) on compacts with a similar total volume fraction of porosity. Compacts of fumed silica are approximately twice as stiff and harder as those of precipitated silica. The behavior of porous silica compacts under indentation is shown to be elastic damageable with a notable influence of the tip geometry on the measured properties. Tests with large spheres are shown to be preferable over sharp tips for estimating the average properties of porous silica compacts.

Author Contributions: Conceptualization and methodology, B.B., C.O. and S.M.; formal analysis, B.B, S.M.; writing—original draft preparation, S.M.; writing—review and editing, G.F., C.O., B.Y. and S.M.; project administration, S.M., C.O., G.F. and B.Y.

Funding: This research was funded by French Ministry for Research, INSA Lyon, and EDF, through the CIFRE grant 2014-0903.

Conflicts of Interest: The authors declare no conflict of interest.

References

1. Bosseboeuf, D. Energy Efficiency Trends and Policies in the Household and Tertiary Sectors. 2015, p. 97. Available online: http://www.odyssee-mure.eu/publications/br/energy-efficiency-trends-policies-buildings.pdf (accessed on 11 March 2019).
2. Intergovernmental Panel on Climate Change. Summary for policymakers. In *Climate Change 2014 Mitigation of Climate Change*; Cambridge University Press: Cambridge, UK, 2015; pp. 1–30.
3. Baetens, R.; Jelle, B.P.; Thue, J.; Tenpierik, M.J.; Grynning, S.; Uvsløkk, S.; Gustavsen, A. Vacuum insulation panels for building applications: A review and beyond. *Energy Build.* **2010**, *42*, 147–172. [CrossRef]
4. Quenard, S.H.D. Micro-nano porous materials for high performance thermal insulation. In Proceedings of the 2nd International Symposium on Nanotechnology in Construction, Bilbao, Spain, 13–16 November 2005.
5. Chal, B.; Yrieix, B.; Roiban, L.; Masenelli-Varlot, K.; Chenal, J.-M.; Foray, G. Nanostructured silica used in super-insulation materials (SIM), hygrothermal ageing followed by sorption characterizations. *Energy Build.* **2019**, *183*, 626–638. [CrossRef]
6. Caps, R.; Fricke, J. Thermal Conductivity of Opacified Powder Filler Materials for Vacuum Insulations1. *Int. J. Thermophys.* **2000**, *21*, 445–452. [CrossRef]
7. Simmler, H.; Heinemann, U.; Kumaran, K.; Brunner, S. Vacuum Insulation Panels: Study on VIP-Components and Panels for Service Life Prediction of VIP in Building Applications Subtask A. Available online: http://www.iea-ebc.org/Data/publications/EBC_Annex_39_Report_Subtask-A.pdf (accessed on 11 March 2019).
8. Baeza, G.P.; Genix, A.-C.; Paupy-Peyronnet, N.; Degrandcourt, C.; Couty, M.; Oberdisse, J. Revealing nanocomposite filler structures by swelling and small-angle X-ray scattering. *Faraday Discuss.* **2016**, *186*, 295–309. [CrossRef] [PubMed]
9. Baeza, G.P.; Genix, A.-C.; Degrandcourt, C.; Petitjean, L.; Gummel, J.; Couty, M.; Oberdisse, J. Multiscale Filler Structure in Simplified Industrial Nanocomposite Silica/SBR Systems Studied by SAXS and TEM. *Macromolecules* **2013**, *46*, 317–329. [CrossRef]
10. Roiban, L.; Foray, G.; Rong, Q.; Perret, A.; Ihiawakrim, D.; Masenelli-Varlot, K.; Mairea, E.; Yrieixb, B. Advanced three dimensional characterization of silica-based ultraporous materials. *RSC Adv.* **2016**, *6*, 10625–10632. [CrossRef]
11. Benane, B.; Baeza, G.P.; Chal, B.; Roiban, L.; Meille, S.; Olagnon, C.; Yrieix, B.; Foray, G. Multiscale Structure of Super Insulation Nano-Fumed Silicas Studied by SAXS, Tomography and Porosimetry. *Acta Mater.* **2019**, *168*, 401–410. [CrossRef]
12. Meille, S.; Garboczi, E.J. Linear elastic properties of 2D and 3D models of porous materials made from elongated objects. *Model. Simul. Mater. Sci. Eng.* **2001**, *9*, 371. [CrossRef]
13. Meille, S.; Lombardi, M.; Chevalier, J.; Montanaro, L. Mechanical properties of porous ceramics in compression: On the transition between elastic, brittle, and cellular behavior. *J. Eur. Ceram. Soc.* **2012**, *32*, 3959–3967. [CrossRef]
14. Deville, S.; Meille, S.; Seuba, J. A meta-analysis of the mechanical properties of ice-templated ceramics and metals. *Sci. Technol. Adv. Mater.* **2015**, *16*, 043501. [CrossRef]
15. Chen, Z.; Wang, X.; Atkinson, A.; Brandon, N. Spherical indentation of porous ceramics: Elasticity and hardness. *J. Eur. Ceram. Soc.* **2016**, *36*, 1435–1445. [CrossRef]
16. Johnson, A.J.W.; Herschler, B.A. A review of the mechanical behavior of CaP and CaP/polymer composites for applications in bone replacement and repair. *Acta Biomater.* **2011**, *7*, 16–30. [CrossRef]
17. Gonzenbach, U.; Studart, A.R. Macroporous Ceramics from Particle-Stabilized Wet Foams. *J. Am. Ceram. Soc.* **2007**, *90*, 16–22. [CrossRef]
18. Pecqueux, F.; Tancret, F.; Payraudeau, N.; Bouler, J.M. Influence of microporosity and macroporosity on the mechanical properties of biphasic calcium phosphate bioceramics: Modelling and experiment. *J. Eur. Ceram. Soc.* **2010**, *30*, 819–829. [CrossRef]
19. Sanahuja, J.; Dormieux, L.; Meille, S.; Hellmich, C.; Fritsch, A. Micromechanical explanation of elasticity and strength of gypsum: From elongated anisotropic crystals to isotropic porous polycrystals. *J. Eng. Mech.* **2010**, *136*. [CrossRef]
20. Moner-Girona, M.; Roig, A.; Molins, E. Micromechanical properties of silica aerogels. *Appl. Phys. Lett.* **1999**, *75*, 653. [CrossRef]

21. Kucheyev, S.O.; Hamza, A.V.; Satcher, J.H.; Worsley, M.A. Depth-sensing indentation of low-density brittle nanoporous solids. *Acta Mater.* **2009**, *57*, 3472–3480. [CrossRef]
22. Alaoui, A.H.; Woignier, T.; Scherer, G.W.; Phalippou, J. Comparison between flexural and uniaxial compression tests to measure the elastic modulus of silica aerogel. *J. Non-Cryst. Solids* **2008**, *354*, 4556–4561. [CrossRef]
23. Woignier, T.; Primera, J.; Alaoui, A.; Etienne, P.; Despestis, F.; Calas-Etienne, S. Mechanical Properties and Brittle Behavior of Silica Aerogels. *Gels* **2015**, *1*, 256–275. [CrossRef]
24. Tillotson, T.M.; Hrubesh, L.W. Transparent ultralow-density silica aerogels prepared by a two-step sol-gel process. *J. Non-Cryst. Solids* **1992**, *145*, 44–50. [CrossRef]
25. Kucheyev, S.O.; Stadermann, M.; Shin, S.J.; Satcher, J.H., Jr.; Gammon, S.A.; Letts, S.A.; van Buuren, T.; Hamza, A.V. Super-compressibility of ultralow-density nanoporous silica. *Adv. Mater.* **2012**, *24*, 776–780. [CrossRef]
26. Kucheyev, S.O.; Lord, K.A.; Hamza, A.V. Room-temperature creep of nanoporous silica. *J. Mater. Res.* **2011**, *26*, 781–784. [CrossRef]
27. Hentzschel, C.M.; Alnaief, M.; Smirnova, I.; Sakmann, A.; Leopold, C.S. Tableting properties of silica aerogel and other silicates. *Drug Dev. Ind. Pharm.* **2012**, *38*, 462–467. [CrossRef]
28. Clément, P.; Meille, S.; Chevalier, J.; Olagnon, C. Mechanical characterization of highly porous inorganic solids materials by instrumented micro-indentation. *Acta Mater.* **2013**, *61*, 6649–6660. [CrossRef]
29. Staub, D.; Meille, S.; Le Corre, V.; Rouleau, L.; Chevalier, J. Identification of a damage criterion of a highly porous alumina ceramic. *Acta Mater.* **2016**, *107*, 261–272. [CrossRef]
30. Gallo, M.; Tadier, S.; Meille, S.; Gremillard, L.; Chevalier, J. The in vitro evolution of resorbable brushite cements: A physico-chemical, micro-structural and mechanical study. *Acta Biomater.* **2017**, *53*, 515–525. [CrossRef]
31. Pathak, S.; Kalidindi, S.R. Spherical nanoindentation stress-strain curves. *Mater. Sci. Eng. R Rep.* **2015**, *91*, 1–36. [CrossRef]
32. Bala, Y.; Depalle, B.; Douillard, T.; Meille, S.; Clément, P.; Follet, H.; Chevalier, J.; Boivin, G. Respective roles of organic and mineral components of human cortical bone matrix in micromechanical behavior: An instrumented indentation study. *J. Mech. Behav. Biomed. Mater.* **2011**, *4*, 7. [CrossRef]
33. Moseson, A.J.; Basu, S.; Barsoum, M.W. Determination of the effective zero point of contact for spherical nanoindentation. *J. Mater. Res.* **2008**, *23*, 204–209. [CrossRef]
34. Flamant, Q.; Caravaca, C.; Meille, S.; Gremillard, L.; Chevalier, J.; Biotteau-Deheuvels, K.; Kuntz, M.; Chandrawati, R.; Herrmann, I.K.; Spicer, C.D.; et al. Selective etching of injection molded zirconia-toughened alumina: Towards osseointegrated and antibacterial ceramic implants. *Acta Biomater.* **2016**, *46*, 308–322. [CrossRef]
35. Oliver, W.C.; Pharr, G.M. An improved technique for determining hardness and elastic modulus using load and displacement sensing indentation experiments. *J. Mater. Res.* **1992**, *7*, 1564–1583. [CrossRef]
36. Lawn, B.R. Indentation of Ceramics with Spheres: A Century after Hertz. *J. Am. Ceram. Soc.* **1998**, *81*, 1977–1994. [CrossRef]
37. Benane, B. *Mechanics of Compacted Silica for Use as VIP (Vacuum Insulation Panels) Core Material*; INSA Lyon: Villeurbanne, France, 2018.
38. Chen, Z.; Wang, X.; Atkinson, A.; Brandon, N. Spherical indentation of porous ceramics: Cracking and toughness. *J. Eur. Ceram. Soc.* **2016**, *36*, 3473–3480. [CrossRef]
39. Schwiedrzik, J.J.; Zysset, P.K. The influence of yield surface shape and damage in the depth-dependent response of bone tissue to nanoindentation using spherical and Berkovich indenters. *Comput. Methods Biomech. Biomed. Eng.* **2013**, *18*, 492–505. [CrossRef]
40. Colombo, P. Materials Science: In Praise of Pores. *Science* **2008**, *322*, 381–383. [CrossRef]
41. Garboczi, E.J.; Snyder, K.A.; Douglas, J.F.; Thorpe, M.F. Geometrical percolation threshold of overlapping ellipsoids. *Phys. Rev. E* **1995**, *52*, 819–828. [CrossRef]
42. Toivola, Y.; Stein, A.; Cook, R.F. Depth-sensing indentation response of ordered silica foam. *J. Mater. Res.* **2004**, *19*, 260–271. [CrossRef]
43. Guesnet, E.; Dendievel, R.; Jauffrès, D.; Martin, C.L.; Yrieix, B. A growth model for the generation of particle aggregates with tunable fractal dimension. *Phys. A Stat. Mech. Appl.* **2019**, *513*, 63–73. [CrossRef]

44. Gonçalves, W.; Amodeo, J.; Morthomas, J.; Chantrenne, P.; Perez, M.; Foray, G.; Martin, C.L. Nanocompression of secondary particles of silica aerogel. *Scr. Mater.* **2018**, *157*, 157–161. [CrossRef]
45. Kermouche, G.; Barthel, E.; Vandembroucq, D.; Dubujet, P. Mechanical modelling of indentation-induced densification in amorphous silica. *Acta Mater.* **2008**, *56*, 3222–3228. [CrossRef]
46. Chen, Z.; Wang, X.; Brandon, N.; Atkinson, A. Analysis of spherical indentation of porous ceramic films. *J. Eur. Ceram. Soc.* **2016**, *37*, 1031–1038. [CrossRef]
47. Hamdi, B.; Gottschalk-Gaudig, T.; Balard, H.; Brendlé, E.; Nedjari, N.; Donnet, J.-B. Ageing process of some pyrogenic silica samples exposed to controlled relative humidities: Part I: Kinetic of water sorption and evolution of the surface silanol density. *Colloids Surf. A Physicochem. Eng. Asp.* **2016**, *491*, 62–69. [CrossRef]
48. Morel, B.; Autissier, L.; Autissier, D.; Lemordant, D.; Yrieix, B.; Quenard, D. Pyrogenic silica ageing under humid atmosphere. *Powder Technol.* **2009**, *190*, 225–229. [CrossRef]
49. Yrieix, B.; Morel, B.; Pons, E. VIP service life assessment: Interactions between barrier laminates and core material, and significance of silica core ageing. *Energy Build.* **2014**, *85*, 617–630. [CrossRef]

© 2019 by the authors. Licensee MDPI, Basel, Switzerland. This article is an open access article distributed under the terms and conditions of the Creative Commons Attribution (CC BY) license (http://creativecommons.org/licenses/by/4.0/).

Article

Simulating Fiber-Reinforced Concrete Mechanical Performance Using CT-Based Fiber Orientation Data

Vladimir Buljak [1,*], Tyler Oesch [2] and Giovanni Bruno [2,3]

[1] Faculty of Mechanical Engineering, University of Belgrade, Kraljice Marije 16, 11120 Belgrade 35, Serbia
[2] Bundesanstalt für Materialforschung und–prüfung, BAM (Federal Institute for Materials Research and Testing), 12205 Berlin, Germany; Tyler.Oesch@bam.de (T.O.); Giovanni.Bruno@bam.de (G.B.)
[3] Institute of Physics and Astronomy, University of Potsdam, Karl-Liebknecht-Str.24-25, 14476 Potsdam, Germany
* Correspondence: vladimir.buljak@polimi.it

Received: 15 January 2019; Accepted: 20 February 2019; Published: 1 March 2019

Abstract: The main hindrance to realistic models of fiber-reinforced concrete (FRC) is the local materials property variation, which does not yet reliably allow simulations at the structural level. The idea presented in this paper makes use of an existing constitutive model, but resolves the problem of localized material variation through X-ray computed tomography (CT)-based pre-processing. First, a three-point bending test of a notched beam is considered, where pre-test fiber orientations are measured using CT. A numerical model is then built with the zone subjected to progressive damage, modeled using an orthotropic damage model. To each of the finite elements within this zone, a local coordinate system is assigned, with its longitudinal direction defined by local fiber orientations. Second, the parameters of the constitutive damage model are determined through inverse analysis using load-displacement data obtained from the test. These parameters are considered to clearly explain the material behavior for any arbitrary external action and fiber orientation, for the same geometrical properties and volumetric ratio of fibers. Third, the effectiveness of the resulting model is demonstrated using a second, "control" experiment. The results of the "control" experiment analyzed in this research compare well with the model results. The ultimate strength was predicted with an error of about 6%, while the work-of-load was predicted within 4%. It demonstrates the potential of this method for accurately predicting the mechanical performance of FRC components.

Keywords: Fiber-reinforced concrete; X-ray computed tomography (CT); anisotropic fiber orientation; inverse analysis

---

1. Introduction

With recent developments in structural and material mechanics, assessments of safety margin with respect to non-linear system response and failure, instead of admissible stresses, became possible and even required by several codes [1–3]. Numerical methods for the accurate simulation of the non-linear behavior of engineering structures have also been developed in last few decades and incorporated into computational tools [4]. This evolution has significantly increased the need for knowledge about the inelastic properties of materials (e.g., plasticity, damage, creep, fracture, etc.) which cannot be assessed, unlike the elastic parameters, by means of non-destructive tests such as those based on ultrasound tests [5]. Furthermore, the assessment of inelastic properties when combined phenomena take place (e.g., plasticity with damage and fracture) is rather difficult, or even impossible, using standardized tests for the evaluation of compressive or tensile strength as a single material property [6,7].

Accurate numerical modeling within the non-linear regime is related to the appropriate selection of a constitutive model, capable of accounting for phenomena that are taking place at the material level (e.g., plastic deformation, damage of the material, creep, etc.). Such a constitutive model would offer a framework for the accurate modeling of a structural response in the general context, beyond the one represented by the experiment performed for its calibration. Therefore, the quantification of the parameters that govern the constitutive equations should not be merely reduced to the fitting of a single experimental response.

The importance of appropriate constitutive model selection becomes more evident when a complex material, such as fiber-reinforced concrete (FRC), should be modeled. Owing to the presence of small fibers, the structural response of FRC with respect to conventional reinforced concrete is considerably different. With conventional reinforcement, significant elongation of the steel bar is required, so that it can carry tensile loads, which requires the notable opening of macro cracks within the concrete. In contrast, in FRC the cracks are often barely visible to the naked eye, and are developed in the form of a network, which gives the structural member greater ductility and, at the same time, limits the exposure of fibers to the ambient conditions [8,9].

In previous years, considerable research efforts have been devoted to studying the mechanical response of FRC. Significant attention has been devoted to analyzing the influence of fiber orientation on the mechanical response of structures [10,11]. Since it is recognized that fiber distribution and orientation play an important role in global mechanical properties, several authors have discussed the influence of the casting process on the orientation of fibers [12], analyzed various methods to measure it [13], and tried to predict it through flow simulations [14]. There have also been many experimental studies focused on the quantification of the global mechanical properties of structural components made of FRC [15–17]. However, for the systematic incorporation of this material into structural analysis, a proper constitutive description and related parameter calibration is required.

The mechanical response of the structural components made of FRC depends, to a large extent, on the local distribution and orientation of reinforcing fibers. Such information can be collected through the use of X-ray computer tomography (CT), but its effective incorporation into numerical modeling still needs to be solved. The major difficulty for successful modeling is related to the fact that the existing orthotropic constitutive damage models, which are implemented in commercial finite element codes, are suitable for defining anisotropic material behavior only at the structural level. While this can be an appropriate strategy to model conventional reinforced concrete, it is not appropriate for the FRC, where locally strengthened directions, achieved by reinforcing fibers, vary considerably within different regions of individual structural components.

For reliable numerical simulations, fiber distribution and orientation should be included within the constitutive description, thus requiring multi-scale approaches with the capability of incorporating the inherent variability of the internal structure. For this purpose, discrete models [18,19], can be used, with further modifications, to take into account fiber distribution and orientations. These discrete models are capable of addressing material behavior at the micro and meso-scales. For the macro-scale, however, which is of importance for the analysis of large-scale structures, it is desirable to have a continuum phenomenological model. Such models are based on a representative volume element (RVE), treated as a continuum, without the necessity to model smaller constituents (e.g., fibers or grains) [20]. The presence of these individual constituents is, instead, taken into account through homogenized, macro-scale mechanical characteristics. These models, necessarily, involve certain assumptions that could limit their applicability. The feasibility of the numerical implementation, however, is significantly improved, since the problem is solved on a single scale. This approach is adopted in the present study.

Considering the nature of the phenomena that take place on the fiber scale, a reasonable approach would be to employ a damage model. The major difficulty related to the employment of existing constitutive damage models within commercial finite element modeling (FEM) codes is that even though they can simulate either isotropic or orthotropic behavior, the orthotropic behavior can only be modeled along the directions defined at the structural level. This can be an appropriate strategy to model conventional reinforced concrete, where reinforcing bars have well established directions with respect to the structure, but is not sufficient for FRC, where locally strengthened directions change from one point to another.

A detailed description of the FRC material as well as the characteristics of the CT measurements and the analysis of fiber orientation data are presented in Section 2. Section 3 is devoted to a description of the main features of the finite element model used in this study and its specialized adaptation to take into account the local material variability with reference to the three-point bending experiment of a notched beam, which was adopted as the test for subsequent material calibration. The inverse analysis procedure developed to assess governing constitutive parameters is outlined in Section 4. Section 4 presents the results produced by the calibrated constitutive model including those from its validation. Section 5 concludes the paper by offering a brief summary of the advantages and limitations of the proposed strategy together with some future prospects.

2. Materials, Computed Tomography Measurements and Data Reconstruction

2.1. Material and Specimen Properties

The FRC material used during this research program is an ultra-high-performance concrete (UHPC) developed by the US Army Engineer Research and Development Center (ERDC, Vicksburg, MS, USA). As a reactive powder concrete mixture, this material contains no course aggregate particles and is characterized by a low water-to-cement ratio and high cement paste content [21–23]. This material has a nominal compressive strength of 200 MPa and contains nominally 3.6% steel fibers by volume [24]. The steel fibers were 30 mm long, 0.55 mm in diameter, and had hook-like pre-deformations at each end.

The two beams analyzed in this study were nominally 220 mm long, 48 mm high, and 30 mm wide. After curing, nominally 18 mm deep by 5 mm wide "notches" were saw cut at the bottom of each beam midway along the length. Given the large aspect-ratio of the beams and the relatively small height and width dimensions relative to fiber length, it was expected that significant anisotropies in fiber orientation could occur, in particular with the fibers oriented principally in the direction of the beam length. Following CT scanning, the specimens were loaded to failure in a three-point bending configuration. Further detail about the materials, specimens, and mechanical testing was reported by Williams, Roth, Trainor, et al. [22–25].

2.2. CT Measurement and Analysis

Prior to mechanical testing, the central section of each beam was scanned using CT; for details of the specific scanning procedure and parameters [24]. Three-dimensional reconstruction of the CT images was completed using a proprietary algorithm provided by the CT instrument vendor (Northstar Imaging, Inc., Rogers, MN, USA)

Analysis of fiber orientation, including the compilation of fiber data into a finite element mesh, was completed using the Fiber Composite Material Analysis module of VGSTUDIO MAX [26]. A section of the CT image from a beam with overlaid finite element mesh can be seen in Figure 1.

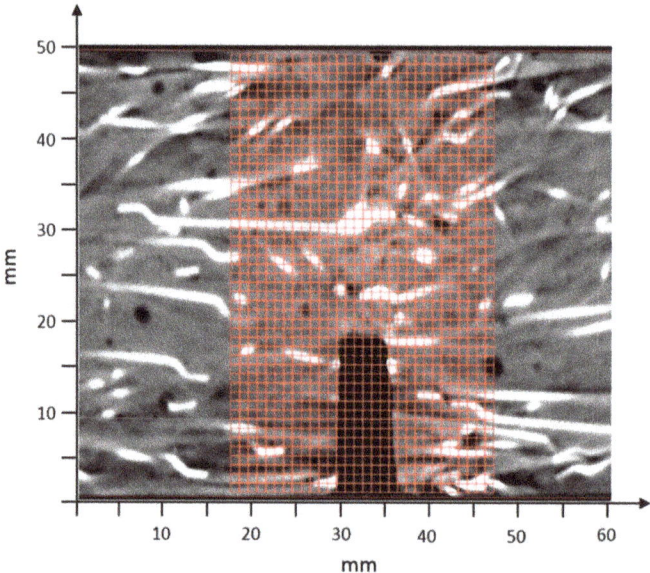

Figure 1. Section of beam computed tomography (CT) image with overlaid finite element mesh.

2.3. Tensor-Based Analysis of Fiber Orientation

In order to achieve a compact description of fiber orientation, a tensorial representation was adopted. Specifically, for a unit vector $\mathbf{p} = [p_1, p_2, p_3]$, the second-order orientation tensor was constructed as its dyadic product [27] namely:

$$\mathbf{T} = \begin{bmatrix} a_{11} & a_{12} & a_{13} \\ a_{21} & a_{22} & a_{23} \\ a_{31} & a_{32} & a_{33} \end{bmatrix} = \begin{bmatrix} p_1^2 & p_1 p_2 & p_1 p_3 \\ p_2 p_1 & p_2^2 & p_2 p_3 \\ p_3 p_1 & p_3 p_2 & p_3^2 \end{bmatrix} \quad (1)$$

The eigenvalue analysis of matrix $\mathbf{T}$ recovers the unit vector $\mathbf{p}$, which is the eigenvector corresponding to the largest eigenvalue. In this case, this is the only eigenvalue larger than zero (and equal to 1). For a group of N fibers, each is associated with a fiber unit vector $\mathbf{p}^i$, i=1, ... N describing its orientation. The average orientation tensor is computed by taking the mean value of entries from the individual orientation tensors, calculated for each $\mathbf{p}^i$ [27], namely:

$$\mathbf{T}^{avg} = \begin{bmatrix} a_{11}^{avg} & a_{12}^{avg} & a_{13}^{avg} \\ a_{21}^{avg} & a_{22}^{avg} & a_{23}^{avg} \\ a_{31}^{avg} & a_{32}^{avg} & a_{33}^{avg} \end{bmatrix}, \text{ with } a_{ij}^{avg} = \frac{1}{N}\sum_{k=1}^{N} a_{ij}^k \quad (2)$$

By performing the eigenvalue analysis of $\mathbf{T}^{avg}$, a set of three eigenvectors and corresponding eigenvalues is obtained. For an arbitrary case, when fibers from the group are not all parallel, all the three eigenvalues are different from zero, while their summation is equal to one. The magnitude of each eigenvalue gives the statistical proportion of the fibers from the analyzed group, aligned along the corresponding eigenvector. Thus, by averaging the orientation of the analyzed fiber set, the smallest error is introduced if the representative direction is taken to be the eigenvector corresponding to the largest eigenvalue. This conclusion stems directly from proper orthogonal decomposition (POD) theory [28,29]. Within POD, it is demonstrated that when representing a set of vectors $\mathbf{p}^i = [p_1^i, p_2^i, p_3^i]^T$

with one component approximation, the error of approximation is minimized if they are projected to the direction of the eigenvector corresponding to the largest eigenvalue of matrix **D**:

$$\mathbf{D} = \begin{bmatrix} d_{11} & d_{12} & d_{13} \\ d_{21} & d_{22} & d_{23} \\ d_{31} & d_{32} & d_{33} \end{bmatrix} = \begin{bmatrix} p_1^1 & p_1^2 & \cdots & p_1^N \\ p_2^1 & p_2^2 & \cdots & p_2^N \\ p_3^1 & p_3^2 & \cdots & p_3^N \end{bmatrix} \cdot \begin{bmatrix} p_1^1 & p_2^1 & p_3^1 \\ p_1^2 & p_2^2 & p_3^2 \\ \cdots & \cdots & \cdots \\ p_1^N & p_2^N & p_3^N \end{bmatrix} \quad (3)$$

It is easily demonstrated that matrices **D** and $\mathbf{T}^{avg}$ are related through a scaling factor $1/N$, since $a_{ij}^{avg} = \frac{1}{N} \sum_{i=1}^{N} p_i^k p_j^k$, while $d_{ij}^{avg} = \sum_{i=1}^{N} p_i^k p_j^k$, and, hence, have the same eigenvectors.

The developed numerical model uses two-dimensional domains. Volume cells, over which fiber orientation is analyzed, are of the size 1 mm × 1 mm in the modeled plane and have a 30 mm thickness. Fibers crossing this volume are averaged according to the above outlined procedure, and replaced by the first two components of the first eigenvector (i.e., projected to modeling plane x-y), see Figure 2.

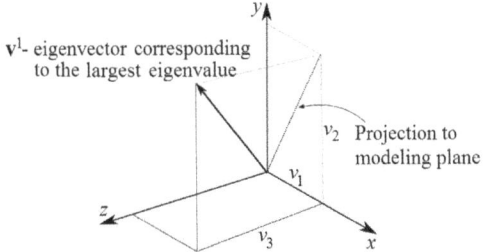

Figure 2. Eigenvector of principal fiber orientation.

The averaging of fiber properties over the 30 mm cell thickness, corresponding to the width of the entire beam, is the required step for implementing the real, 3D fiber measurements into the 2D numerical model.

3. Modeling

3.1. Constitutive Model for Progressive Damage of Fiber-Reinforced Concrete

In this paper, a fully phenomenological orthotropic damage model is adopted to model FRC (i.e., a model that accounts for the effect of fibers and not their physical presence). The presence of fibers contributes to the increase in load-carrying capacity of the whole specimen. This increase is manifested only along the direction of fibers, while in the perpendicular directions the response is considered to roughly correspond to that of unreinforced concrete (though possibly somewhat weaker). Therefore, to model the response employing the phenomenological constitutive model, it is appropriate to use the orthotropic mode: a model with different behavior along different, well-defined directions. The local variability of the structure, in terms of fiber orientation, is incorporated by previous pre-processing. This is achieved by dividing the portion of the sample most susceptible to progressive damage, due to formation of a crack network, into 1 mm × 1 mm 2D elements. Within each of these regions, local coordinate systems are assigned based on fiber-orientation data collected from CT measurements, averaged over the considered region. Thus, through the use of local coordinate systems, the primary direction strengthened by the reinforcement of the fibers in each individual zone is simulated. The constitutive parameters are the same for the whole sample, while local structural variability is accounted for through specific coordinate systems. This formulation results in a constitutive model capable of predicting unique global specimen behavior, since the variability at the structural level, including significant differences in overall structural response, result from changes in these local orientations. Such an approach provides the advantage that the problem is solved over one scale only, requiring the quantification of only one set of constitutive parameters, here solved on the basis of the designated inverse analysis procedure.

The adopted constitutive model is typical for fiber-reinforced composites, existing in the commercial FEM code ABAQUS (Dassault systemes, Providence, RI, USA) [4]. These materials usually exhibit elastic-brittle behavior, with the damage initiation without any previous plastic deformation. Damage here refers to the onset of degradation at a material point, implemented within the continuum constitutive model through the reduction of elastic constants. Given the significant difference in the reinforcing behavior along the axis of the fiber compared to the reinforcing behavior in the direction perpendicular to this axis, an orthotropic damage model is adopted. Within this model, stresses are related to the total strains through following relation:

$$\sigma = \mathbf{C}_D \cdot \varepsilon \qquad (4)$$

with the elasticity matrix computed by:

$$\mathbf{C}_D = \frac{1}{D} \begin{bmatrix} (1-d_f)E_1 & (1-d_f)(1-d_m)\nu_{21}E_1 & 0 \\ (1-d_f)(1-d_m)\nu_{12}E_2 & (1-d_f)E_2 & 0 \\ 0 & 0 & (1-d_s)G \cdot D \end{bmatrix} \qquad (5)$$

where D is calculated as:

$$D = 1 - \left(1 - d_f\right)(1 - d_m)\nu_{12}\nu_{21} \qquad (6)$$

The material mechanical response is therefore governed by the following parameters:

- E_1—Young's modulus for longitudinal direction (i.e., along axis of the fiber)
- E_2—Young's modulus for transversal direction (i.e., corresponding to unreinforced concrete properties)
- ν_{12}, ν_{21}—Poisson's ratios
- d_f, d_m and d_s—Damage variables for the longitudinal load capacity, the transversal load capacity and for shear capacity, respectively

In the general version of the model, the two Young's moduli are not the same, however within this study, the model is simplified by assuming the same value for both moduli. The damage parameters are bounded by the initial value of zero, corresponding to "virgin" material, and the maximum value of one, corresponding to fully degraded material. The calculation of each particular damage parameter value was related to the formulation of the damage initiation criterion (i.e., the state of stress at which the damage parameter value begins to increase above zero) and the criterion by which the damage parameter evolves up to a value of one, representing complete failure. This second damage evolution criterion completely governed the post-damage behavior of the material in the model.

The model adopted in this study (outlined above) required, besides the elastic parameters, the definition of the following parameters governing the damage:

- Damage initiation relative to the longitudinal tensile capacity (and, thus, an increase in d_f) was assumed to occur once the longitudinal tensile stress exceeds the predefined value of σ_f, corresponding to the beginning of fiber reinforcement failure. Here, fiber reinforcement failure is taken to mean the point at which the load carrying capacity of the fibers begins to decrease. This may be due to one or more phenomena (for instance excessive plastic deformation of the fibers, slip along the fiber-mortar interface, etc.);
- Damage initiation relative to the transverse tensile capacity (and, thus, an increase in d_m) was assumed to occur once the transverse tensile stress exceeded the predefined value of σ_c, corresponding to the beginning of unreinforced concrete tensile failure. In other words, fibers are assumed not to contribute to strengthening in the direction perpendicular to their long axis;

- Damage initiation relative to compression capacity is defined identically for both the longitudinal and transverse load capacities (corresponding to an increase in d_f or d_m, respectively). This initiation was assumed to occur once the compressive stress in a given direction exceeded the predefined value of σ_{cmp}, corresponding to the beginning of unreinforced concrete compression failure. This simplified damage initiation criterion relies on the assumption that there is no significant contribution of fibers relative to compression failure;
- Damage initiation relative to the shear capacity (corresponding to an increase in d_s) was assumed to occur once the shear stress exceeds the predefined value of σ_{SL} or σ_{ST}, corresponding to the beginning of concrete shear failure in the longitudinal or transversal directions, respectively.
- It was assumed that the damage propagation was to be governed by the fracture energy dissipated up to the full degradation of the element. In the general version of the model, separate fracture energy parameters governed the damage propagation in longitudinal (d_f), transversal (d_m) and shear (d_s) loading. In this study, however, in view of the 2D (i.e., simplified) nature of the model, a simplification was adopted by setting the value of each of these separate fracture energy parameters equal to the value of a single parameter, G_f. Within the 2D model, it is not possible to model the complex crack pattern that is developed within the real, 3D sample. To compensate for this limitation of the model, it is expected that the G_f parameter will be overestimated. Moreover, since it was expected that the response of the beam was to be dominated by the damage along the longitudinal direction of the fibers, the values of the fracture energy parameters for transversal and shear loading were considered of secondary importance to overall response. The effects of restricting all fracture energy parameters to a single value on the overall beam response was, consequently, considered to be negligible.

3.2. Finite Element Model of Three-Point Bending Test

The geometry of the three-point bending specimen considered in this model was adopted [24], specifically a beam with 48 mm × 30 mm rectangular cross-section and 220 mm overall length, with 200 mm of span between the supports. At the mid-span, below the point of load application, a notch 5 mm wide and 18 mm in depth was introduced into the beam. In order to significantly reduce the necessary computation time, a two-dimensional numerical model was implemented. The beam was modeled as a 2D plane-stress problem, while the cylinders, through which the loading was applied, were considered rigid analytical curves in unilateral contact with the deformable specimen. Between the specimen and the cylinders, contact with friction was assumed, with a Coulomb friction coefficient equal to 0.15, taken as an a priori known quantity. During a three-point bending test, the contact between the specimen and the supports does not exhibit significant sliding. The friction, therefore, does not have a large influence on the response and, thus, an ad hoc value suggested by the literature was assumed [30]. The modeled beam was divided into three zones. In the central zone, which was subjected to the largest value of bending moment, the orthotropic damage model described in Section 3.1 was implemented. In the two outer zones, which were subjected to far lower bending moments, a linear-elastic model was implemented (see Figure 3). The central zone included all material within 10 mm of each side of the notch, thus having the overall width of 25 mm. Due to the load pattern and the specific geometry of the considered specimen, it was expected that the cracks would be formed mostly within this zone. The finite element mesh was generated using the automatic advancing front algorithm in ABAQUS [4], which resulted in a somewhat different shape of the mesh in the two outer zones. However, these two regions of the beam, which are within the elastic range, exhibit rather small deformations and, therefore, do not significantly affect the solution. The finite element mesh selected for the model was verified through a usual procedure by comparing the results of the simulations of models with different mesh densities. Here specifically, the adopted model is compared against one with a significantly denser mesh (having an overall number of degrees of freedom (DOF) about 2.5 times larger). The larger numerical model led to results less than 1% different from those achieved by the adopted model with the coarser mesh. The comparison is not included in the paper for the sake of brevity. In what follows, a brief outline of the adopted damage constitutive model and its specialized adaptation in the present context is given.

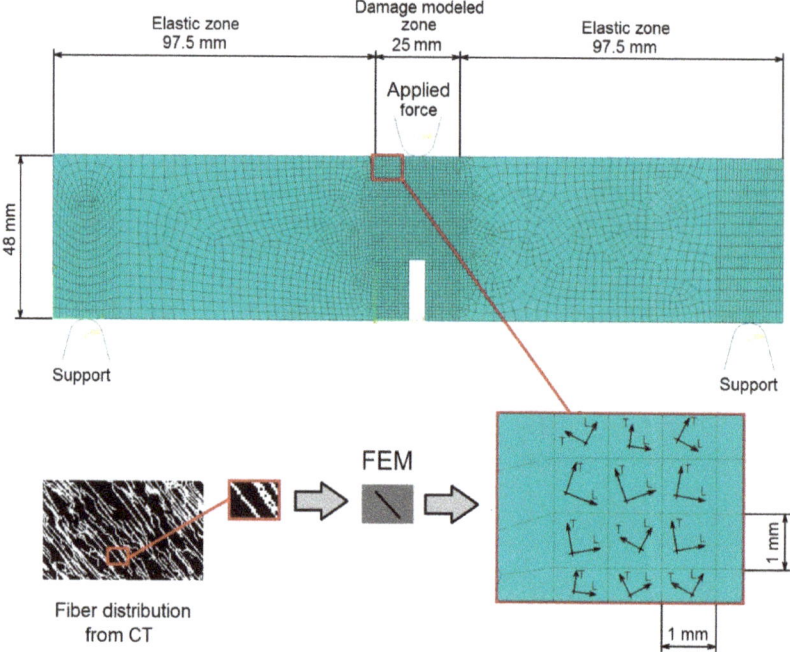

Figure 3. Adopted 2D finite element modeling (FEM) of three-point bending test with individual elements simulated by different constitutive models.

In order to take the local orientation of fibers into account, the central zone of the beam specimen was divided into 1 mm × 1 mm elements, each of these having its own unique local coordinate system. The local coordinate system of these elements was assigned such that its longitudinal (i.e., stronger) axis corresponded to the primary axis of the average orientations of the fibers within the element calculated using CT data from the specimen (see Figure 3 and Section 2.2). For regions without fibers, unreinforced concrete properties were assigned, thus providing the possibility for the model to also partially take into account fiber density distribution. Clearly, this approximation represents a simplification in view of the two-dimensional nature of the model. Variations in material properties (including fiber orientations) over the beam thickness were not considered within the 2D model (i.e., the volume of each element is 1 mm × 1 mm × 30 mm). This strategy served to test the proposed approach prior to its implementation within a more realistic three-dimensional model.

To verify the capability of the proposed approach to model diverse structural responses governed by local variability of fiber distribution, the following numerical exercise was performed. Two different numerical models of three-point bending beams were generated. The fiber distributions were arbitrarily selected to produce a significantly different mechanical response. Such variations in fiber orientation and distribution are commonly seen in FRC and result directly from variation in the material flow during the casting process [11,31]. The numerical simulations lead to the results depicted in Figures 4 and 5. From these figures, it may be observed that a significantly different structural response is obtained in terms of both the overall force-displacement curve and the cracking pattern.

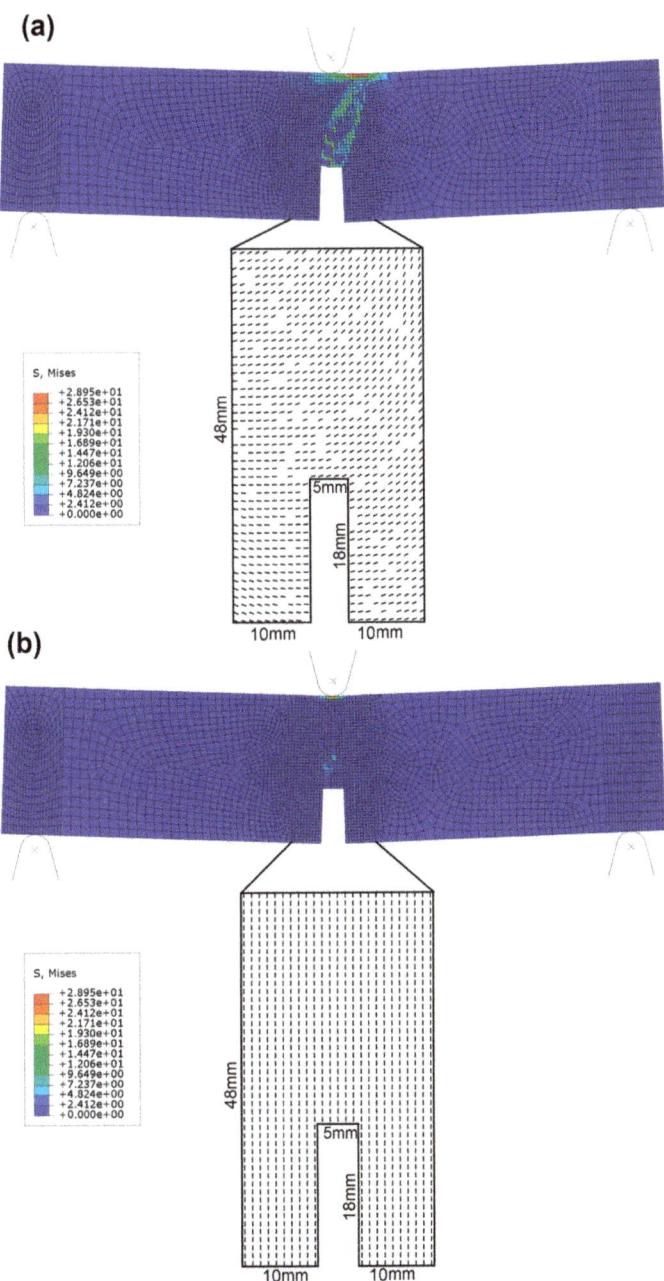

Figure 4. Two numerical models of the three-point bending test with different fiber orientations. (**a**) model with gradually changing fiber orientations, (**b**) model with all fibers oriented perpendicular to the longitudinal axis of the beam.

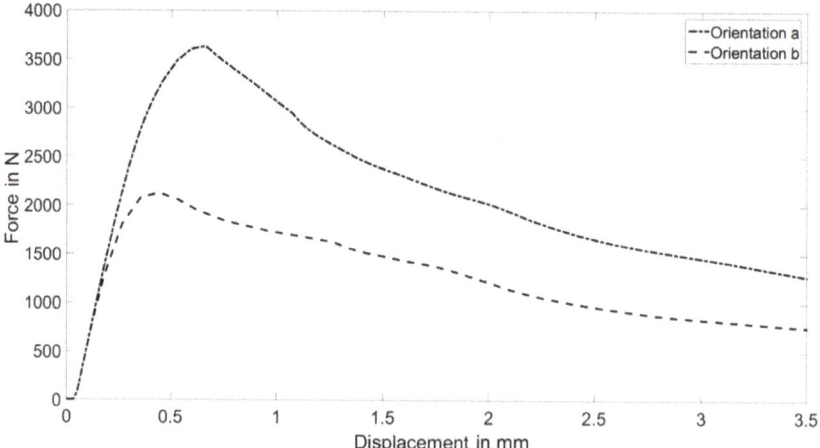

Figure 5. Force displacement curves resulting from the two numerical models of the three-point bending test with different fiber orientations.

Figure 4a depicts the model with gradually changing fiber orientations; Figure 4b depicts the model with all fibers oriented perpendicular to the longitudinal axis of the beam. The dimensions of the whole beam are given in Figure 3.

Numerical model (b) assumed all the fibers to be perpendicular to the longitudinal axis of the beam, therefore providing negligible reinforcing capacity for the three-point-bending load scenario. Indeed, as shown in Figure 5, the maximum force is significantly smaller than for the model (a). Furthermore, such uniform distribution does not provide any variability and, thus, the beam failed with one major crack in the mid-span (see Figure 4b). On the other hand, model (a) assumed that the fibers gradually changed orientation moving from left to right and from the bottom to the top of the sample, with certain zones not containing any fibers. Such a fiber distribution clearly leads to the strengthening of the specimen with respect to the model (b). Additionally, this fiber distribution produced a variation of the locally strengthened directions over the specimen. The effects of this characteristic were captured by the model and can be identified as a network of small, distributed cracks (depicted by green color in Figure 4a).

4. Model Calibration and Validation

In the procedure implemented for this study, a three-point bending test, from which a force-displacement curve is obtained, is treated as the main experimental data. The measured fiber orientation characteristics for the beam used in this calibration are also, naturally, implemented into the numerical model and, thus, directly influence the corresponding constitutive parameter values. A discrepancy function is further constructed to quantify the difference between measured quantities and their computed counterparts [29]. Through the execution of a test simulation employing the above constitutive model, this function becomes dependent on the governing constitutive parameters. At this point, the discrepancy function is minimized with respect to the sought constitutive parameters. This is the solution of the inverse problem.

Afterwards, the assessed parameter values can be treated as material representative data and can be used for arbitrary loading scenarios. Validation of the accuracy and resilience of the model was completed using CT-based fiber orientation measurements from a second three-point bending test in combination with the assessed material parameters to predict the structural response. This is the validation step [32].

4.1. Inverse Analysis Procedure for Quantification of Material Parameters

The reliability of numerical simulations of FRC structural components using the previously outlined damage model rests on the accuracy of the implemented constitutive parameters. In the present case, these parameters quantify the elastic response and progressive damage. In this context, further simplifications were adopted to reduce the number of necessary parameters:

- The values of Young's modulus E_1 (fiber direction) and E_2 (transversal direction) were defined within the model based on experimental data, while the remaining elastic constants were considered to be a priori known values and were, correspondingly, fixed within the model. The response of the beam is dominantly governed by the damage parameters, regardless of the initial, perfectly elastic conditions. Thus, in order to reduce the number of parameters that had to be assessed, only one Young's modulus value (identical for both E_1 and E_2) was used. This simplification is reasonable, considering that the variation in Young's modulus due to the presence of fibers has been previously reported as only about 10% of the Young's modulus typical for unreinforced concrete [33].
- number of damage-initiation parameters were also defined in the model based on experimental data: the damage initiation stress in tension for the longitudinal direction (σ_f), the damage initiation stress in tension for the transversal direction (σ_c), the damage initiation stress in shearing for the longitudinal (σ_{SL}) and for the transversal (σ_{ST}) directions. The stress value for damage initiation in compression was assumed to be the same for both the longitudinal and the transversal directions and did not have to be directly calculated from the experimental data. This damage initiation stress in compression was therefore not subjected to the identification from the experiment and was taken to be eight times the magnitude of the defined damage initiation stresses for tension in the transversal direction, based on standard tensile-to-compression strength ratios described in ACI 318 [34]. This represents a reasonable simplification, assuming that there is no significant contribution of fibers to cracking loads in compression.

Using this formulation, the number of parameters that had to be defined based on experimental data was reduced to six: one elastic parameter, four damage-initiation parameters, and the fracture energy, which was described in Section 3.1. These parameters were quantified through an inverse analysis procedure completed using force-displacement data collected from the three-point-bending experiment. The main features of this inverse analysis procedure are briefly outlined in what follows.

The unknown material parameters were calculated through an inverse analysis procedure in which a suitably selected discrepancy function designed to quantify the difference between measured and numerically computed quantities is minimized. The function has the following form:

$$\omega(\mathbf{p}) = [\mathbf{u}_e - \mathbf{u}_c(\mathbf{p})]^T [\mathbf{u}_e - \mathbf{u}_c(\mathbf{p})] \tag{7}$$

In this equation, p represented a variable vector, here specifically containing the six unknown material parameter variables. The vector $\mathbf{u}_e$ contained the force values at each of N points along the force-displacement curve, corresponding to the equidistant displacements, measured during the experiments. The vector $\mathbf{u}_c$ contains predictions for the force values at each of the N points. These predicted values were generated through test simulations, attributing to the material parameters estimated values corresponding to the current iteration of the optimization algorithm. In Figure 6, differences to be minimized are schematically visualized.

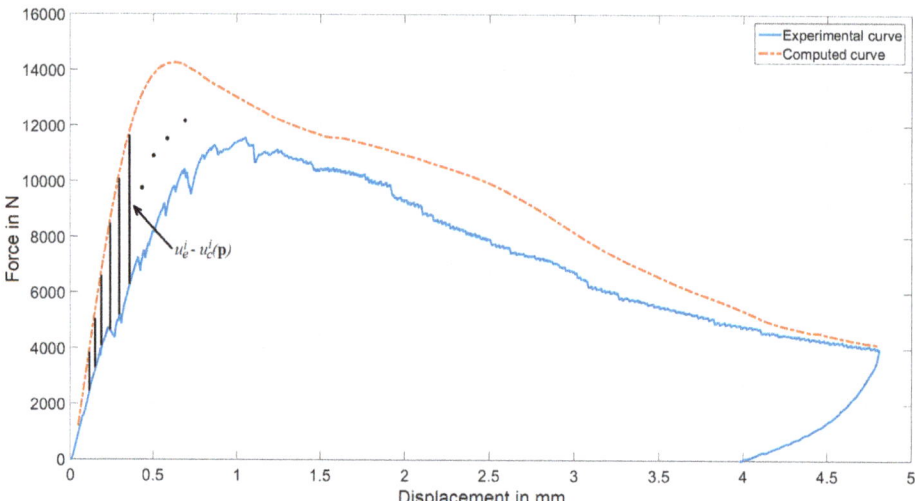

Figure 6. Construction of the discrepancy function from the force-displacement curve.

This function was numerically minimized by employing the "Trust Region" (TR) algorithm. Details of the TR algorithm and its numerical implementation are available in the literature [29,35,36], while in what follows, only the main features are outlined.

The minimization started from an initial vector of the parameters. Within each iteration, a quadratic programming problem in two variable spaces was solved, namely a sub-space spanned by the gradient direction and the second derivative direction (called the Newton direction). The solution of the constrained minimization of this sub-problem provided the modified parameter vector for the next iteration (say $\mathbf{p}_{k+1}$) and resulted in an improved value of the discrepancy function (also called the objective function within the minimization community). Constraints are provided by a "trust region", here adopted as a circle shape within the two-dimensional sub-space.

With such a formulation, the minimization problem was solved inside the trust region, in which the quadratic approximation was "trusted" to be a reasonably good approximation of the real discrepancy function. The quadratic approximation of $\omega(\mathbf{p})$ in the point $\mathbf{p}_k$ reached after the kth iteration was generated by a means of the computed gradient in that point and by means of the Hessian matrix approximated through the Jacobean (**J**), namely:

$$\omega(\mathbf{p}_k + \Delta\mathbf{p}_k) \approx \omega(\mathbf{p}_k) + \Delta\mathbf{p}_k^T \cdot \frac{\partial \omega}{\partial \mathbf{p}}\left(\mathbf{p}^k\right) + \frac{1}{2}\Delta\mathbf{p}_k^T \cdot \mathbf{H} \cdot \Delta\mathbf{p}_k, \text{ where } \mathbf{H} = \mathbf{J}^T\mathbf{J} \qquad (8)$$

The described features of the TR algorithm clearly imply calculations of first derivatives only, numerically computed through finite differences, namely by separately varying each of the parameters. After individual simulations had been completed for variations in each separate parameter variable, a final simulation was completed with the revised values for all parameter variables applied. Therefore, the iterations contain $M+1$ simulations, where M is the number of parameters.

This iterative sequence was repeated until reaching the convergence criteria, here imposed as the change in the discrepancy function value and the difference between parameter vector norms between two consecutive iterations (specifically 1E-2, for the latter, and 1E-4, for the former, criterion). This was realized through an appropriate normalization of parameters with diverse orders of magnitude. The resulting parameter vector reached after a certain number of iterations represented the solution to the designed inverse problem and, therefore, also the representative material properties. As a remedy to the possible lack of convexity of the discrepancy function (7) and consequent termination in a local minimum, the TR iterative sequence was repeated starting from another initialization point.

4.2. Calibration of the Model

The inverse analysis procedure described in Section 4.1 was here employed to assess previously described unknown elastic and damage parameters of the outlined constitutive model. The experimental result used, as an input to the procedure, is a force-displacement curve collected from a three-point bending experiment on the same sample, described in Section 2.1. Based on comparisons of model predictions with this experimental data, a discrepancy function (7) was formed. Additionally, on the basis of CT measurements, the distribution of fiber orientation was calculated for the numerical model. The central zone of the beam, modeled by the damage constitutive model, was discretized by a finite element mesh with square elements 1 mm in size, resulting in the overall number of 1110 elements. To each of these elements, a unique local orientation was attributed in accordance with the procedure described in Section 2.3. This orientation corresponds to the fiber orientations measured in the specimen prior to testing.

The inverse problem was found to be well-posed and the procedure converged to the solution after several iterations. The parameter values, which resulted as the solution to the inverse problem are listed in Table 1. The comparison between experimental and numerical force-displacement curves generated by employing these parameters is visualized in Figure 7. Good agreement between the two curves proves that the adopted approach is capable of capturing the overall mechanical response of three-point bending experiments.

Table 1. Parameter estimates resulting from the inverse analysis.

Parameter	Resulting Value
$E_{1/2}$	26.5 GPa
σ_f	23.6 MPa
σ_c	3.9 MPa
σ_{ST}	2.7 MPa
σ_{SL}	18.3 MPa
G_f	27500 N/m

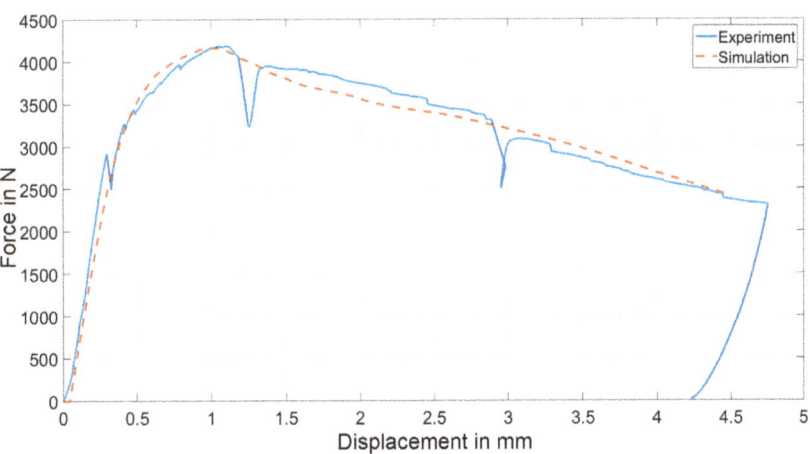

Figure 7. Comparison of force-displacement curves resulting from the experiment and resulting from simulations using inverse analysis.

Considering the adopted simplifications outlined in Section 4.1, the assessed value of Young's modulus represents the mean value of E_1 and E_2. The resulting value of Young's modulus was somewhat smaller than expected. This discrepancy is likely due to the influence of several factors on the displacement values measured during the experiments (e.g., compliance of the beam supports). The elastic properties, however, were not of primary interest here, since they can be measured more precisely with some alternative methods, therefore the removal of these effects was not considered. The damage initiation stresses are of the expected order of magnitude, when compared to the values reported in the literature [14]. The magnitude of parameter σ_f is comparable to flexural strengths previously measured for this material [23], while σ_c corresponds well to the tensile strength of unreinforced concrete. The identification procedure converged to these values without any preconditioning. A comment should also be made here regarding the parameter G_f. The crack network developed in the two-dimensional model is significantly simpler than in the real three-dimensional case. Therefore, the only way for the numerical model to simulate the experimentally observed ductility is by increasing the value of G_f. This could be confirmed by comparing the obtained values of G_f with the absolute values of dissipated energy corresponding to different internal mechanisms [24].

4.3. Validation of the Model

In order to verify the accuracy of the resulting material parameters for the given sample geometry and test setup, a second sample subjected to the three-point bending experiment is further considered. The two samples were made of nominally the same material, but with fairly different fiber orientation and distribution. Another numerical model was built, in which the fiber orientation was distributed within the model following the procedure described in Section 3.2 on the basis of CT measurements of the second beam sample before the test. The constitutive parameters were assumed to have the values resulting from the inverse analysis procedure described above, see Table 1. The experimental and numerical force-displacement curves for this second beam also turned out to be in quite good agreement (see Figure 8). This result corroborates the conclusion that the proposed strategy of modeling FRC by separating the material behavior (here expressed in terms of elastic-damage model) from the local structural variability, accounted for by applying different fiber orientations on an element-wise basis, is quite promising. The two considered specimens were different only in terms of fiber distribution, and two numerical models with the same material properties captured the unique structural response of each beam quite well.

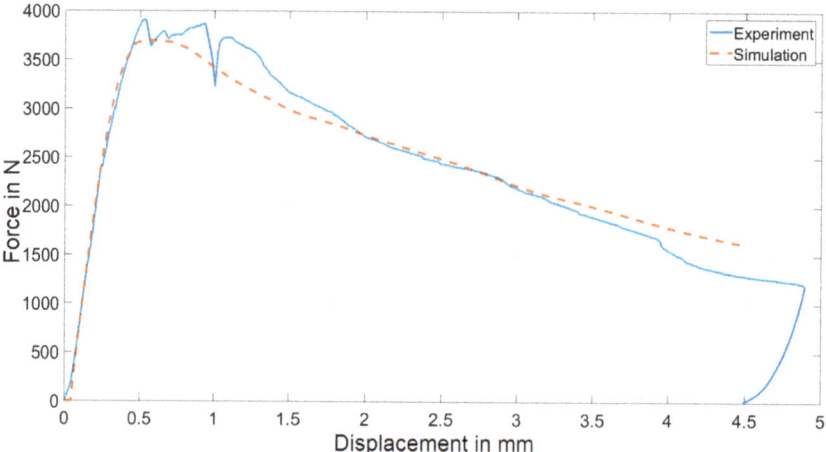

Figure 8. Comparison of experimental and simulated force-displacement curves for the second experiment.

5. Closing Remarks and Future Prospects

This research showed that with a fully phenomenological model, without detailed modeling of fibers, the ductility of the fiber-reinforced concrete could be successfully simulated. Further, it is clear that such behavior can be captured with a single material property set, while the structural variability is incorporated through localized fiber orientation. The novelty of the proposed approach lies in the combination of the phenomenologically based model construction with the inverse analysis based calibration procedure. One obvious limitation of the modeling strategy presented here is the use of a simplified two-dimensional numerical model. The extension of this approach to a three-dimensional model requires the implementation of the proposed constitutive model using the USER subroutine of the commercial software ABAQUS. Such an extension would be a desirable development, but requires further research, which is presently ongoing, and additional numerical implementation that extends the relevant functionalities of the commercial software. The results achieved and presented within this paper, however, provide a meaningful contribution by demonstrating the advantages and limitations of this modeling approach using readily available numerical tools.

Author Contributions: V.B. contributed to conceptualization, formal analysis, software development, writing the original draft and further reviewing and editing. T.O. contributed to conceptualization, CT data analysis, writing the original draft and further reviewing and editing. G.B. contributed to conceptualization, supervision and reviewing and editing.

Funding: This research received no external funding.

Acknowledgments: The authors would like to thank Prof. Eric Landis from the University of Maine for providing the experimental data, including CT images, used in this paper for the calibration and validation of the numerical model. VB acknowledges travel support from BAM.

Conflicts of Interest: The authors declare no conflict of interest.

References

1. European Committee for Standardization. *Eurocode 2: Design of Concrete Structures—Part 1-1: General Rules and Rules for Buildings;* EN 1992-1-1; European Committee for Standardization: Brusseles, Belgium, 2004.
2. BSSC. *NEHRP Guidelines for the Seismic Rehabilitation of Buildings—FEMA Publication 273;* Building Seismic Safety Council: Washington, DC, USA, 1997.
3. *Minimum Design Loads for Buildings and Other Structures—Revision of ASCE 7-98;* American Society of Civil Engineers: Reston, VA, USA, 2006.
4. ABAQUS. *Theory and User's Manuals, Release 6.10;* Assault Systèmes Simulia Corp.: Providence, RI, USA, 2010.
5. Martin, L.P.; Lindgren, E.A.; Rosen, M.; Sidhu, H. Ultrasonic determination of elastic moduli in cement during hydrostatic loading to 1 GPa. *Mater. Sci. Eng. A* **2000**, *279*, 87–95. [CrossRef]
6. Standard Test Method for Compressive Strength of Cylindrical Concrete Specimens. Available online: https://www.astm.org/Standards/C39 (accessed on 15 January 2019).
7. ASTM-C496/C496M. Standard Test Method for Splitting Tensile Strength of Cylindrical Concrete Specimens. Available online: https://www.astm.org/Standards/C496 (accessed on 15 January 2019).
8. Fantilli, A.P.; Mihashi, H.; El-Tawail, S. Tailoring SHCC made of steel cords and plastic fibers. *High Perform. Fiber Reinf. Cem. Compos. 6 HPFRCC 6* **2012**, *2*, 11–18.
9. Stähli, P.; Mier, J.G.M.V. Manufacturing, fibre anisotropy and fracture of hybrid fibre concrete. *Eng. Fract. Mech.* **2007**, *74*, 223–242. [CrossRef]
10. Boulekbache, B.; Hamrat, M.; Chemrouk, M.; Amziane, S. Flowability of fibre-reinforced concrete and its effect on the mechanical properties of the material. *Constr. Build. Mater.* **2010**, *24*, 1664–1671. [CrossRef]
11. Stähli, P.; Custer, R.; Mier, A.G.M.V. On flow properties, fibre distribution, fibre orientation and flexural behaviour of FRC. *Mater. Struct.* **2008**, *41*, 189–196. [CrossRef]
12. Laranjeira, F.; Aguado, A.; Molins, C.; Grünewald, S.; Walraven, J.; Cavalaro, S. Framework to predict the orientation of fibers in FRC: A novel philosophy. *Cem. Concr. Res.* **2012**, *42*, 752–768. [CrossRef]
13. Liu, J.; Sun, W.; Miao, C.; Liu, J.; Li, C. Assessment of fiber distribution in steel fiber mortar using image analysis. *J. Wuhan Univ. Technol. Mater. Sci. Ed.* **2012**, *27*, 166–171. [CrossRef]

14. Švec, O.; Zirgulis, G.; Bolander, J.E.; Stang, H. Influence of formwork surface on the orientation of steel fibres within self-compacting concrete and on the mechanical properties of cast structural elements. *Cem. Concr. Compos.* **2014**, *50*, 60–72. [CrossRef]
15. Molins, C.; Aguado, A.; Saludes, S. Double Punch Test to control the energy dissipation in tension of FRC (Barcelona test). *Mater. Struct.* **2009**, *42*, 415–425. [CrossRef]
16. Di Prisco, M.; Ferrara, L.; Lamperti, M.G. Double edge wedge splitting (DEWS): An indirect tension test to identify post-cracking behaviour of fibre reinforced cementitious composites. *Mater. Struct.* **2013**, *46*, 1893–1918. [CrossRef]
17. Zain, M.F.M.; Mahmud, H.B.; deIlhama, A.; Faizala, M. Prediction of splitting tensile strength of high-performance concrete. *Cem. Concr. Res.* **2002**, *32*, 1251–1258. [CrossRef]
18. Ibrahimbegovic, A.; Delaplace, A. Microscale and mesoscale discrete models for dynamic fracture of structures built of brittle material. *Comput. Struct.* **2003**, *81*, 1255–1270. [CrossRef]
19. Cusatis, G.; Pelessone, D.; Mencarelli, A. Lattice Discrete Particle Model (LDPM) for failure behavior of concrete. I: Theory. *Cem. Concr. Compos.* **2011**, *33*, 11–20. [CrossRef]
20. Drugan, W.J.; Willis, J.R. A micromechanics-based nonlocal constitutive equation and estimates of representative volume element size for elastic composites. *J. Mech. Phys. Solids* **1996**, *44*, 497–524. [CrossRef]
21. Bonneau, O.; Poulin, C.; Dugat, J.; Richard, P.; Aitcin, P. Reactive powder concrete: From theory to practice. *Concr. Int.* **1996**, *18*, 47–49.
22. Williams, E.M.; Graham, S.S.; Reed, P.A.; Rushing, T.S. *Laboratory Characterization of Cor-Tuf Concrete With and Without Steel Fibers*, ERDC/GSL ed.; Army, U.S., Ed.; Engineer Research and Development Center: Vicksburg, MS, USA, 2009.
23. Roth, J.M.; Rushing, T.S.; Flores, O.G.; Sham, D.K.; Stevens, J.W. *Laboratory Investigation of the Characterization of Cor-Tuf Flexural and Splitting Tensile Properties*, ERDC/GSL ed.; Army, U.S., Ed.; Engineer Research and Development Center: Vicksburg, MS, USA, 2010.
24. Trainor, K.J.; Foust, B.W.; Landis, E.N. Measurement of Energy Dissipation Mechanisms in Fracture of Fiber-Reinforced Ultrahigh-Strength Cement-Based Composites. *J. Eng. Mech.* **2013**, *139*, 771–779. [CrossRef]
25. Trainor, K.J. *3-D Analysis of Energy Dissipation Mechanisms in Steel Fiber Reinforced Reactive Powder Concrete*; University of Maine: Orono, ME, USA, 2011.
26. *VGSTUDIO MAX*, version 3.2; Volume Graphics GmbH: Heidelberg, Germany, 2018.
27. Advani, S.; Tucker, C. The use of tensors to describe and predict fiber orientation in short fiber composites. *J. Rheol.* **1987**, *31*, 751–784. [CrossRef]
28. Buljak, V.; Maier, G. Proper orthogonal decomposition and radial basis functions in material characterization based on instrumented indentation. *Eng. Struct.* **2011**, *33*, 492–501. [CrossRef]
29. Buljak, V. *Inverse Analysis with Model Reduction: Proper Orthogonal Decomposition in Structural Mechanics*; Springer: Berlin, Germany, 2011.
30. Pallett, P.; Gorst, N.; Clark, L.; Thomas, D. Friction resistance in temporary works materials. *Concrete* **2002**, *36*, 12–15.
31. Oesch, T.; Landis, E.; Kuchma, D. A methodology for quantifying the impact of castic procedure on anisotropy in fiber-reinforced concrete using X-ray CT. *Mater. Struct.* **2018**, *51*, 73. [CrossRef]
32. Buljak, V.; Bruno, G. Numerical modeling of thermally induced microcracking in porous ceramics: An approach using cohesive elements. *J. Eur. Ceram. Soc.* **2018**, *38*, 4099–4108. [CrossRef]
33. Gul, M.; Bashir, A.; Naqash, J.A. Study of modulus of elasticity of steel fiber reinforced concrete. *Int. J. Eng. Adv. Technol.* **2014**, *3*, 304–309.
34. *Building Code Requirements for Structural Concrete (ACI 318-14)*; American Concrete Institute: Farmington Hills, MI, USA, 2014.
35. Buljak, V.; Bocciarelli, M.; Maier, G. Mechanical characterization of anisotropic elasto-plastic materials by indentation curves only. *Meccanica* **2014**, *49*, 1587–1599. [CrossRef]
36. Nocedal, J.; Wright, S.J. *Numerical Optimization*; Springer: New York, NY, USA, 2006.

© 2019 by the authors. Licensee MDPI, Basel, Switzerland. This article is an open access article distributed under the terms and conditions of the Creative Commons Attribution (CC BY) license (http://creativecommons.org/licenses/by/4.0/).

Article

Influence of Activator Na₂O Concentration on Residual Strengths of Alkali-Activated Slag Mortar upon Exposure to Elevated Temperatures

Tai Thanh Tran and Hyug-Moon Kwon *

Department of Civil Engineering, Yeungnam University, Gyeongsan, Gyeongbuk 712-749, Korea; thanhtaivlxd@gmail.com
* Correspondence: hmkwon@yu.ac.kr; Tel.: +82-053-810-2411

Received: 15 June 2018; Accepted: 24 July 2018; Published: 27 July 2018

Abstract: The mechanical strength variation of ambient cured Alkali-activated mortar (AAS) upon exposure to elevated temperatures from 200 to 1200 °C was studied in this article. Slag was activated by the combination of sodium silicate liquid (Na_2SiO_3) and sodium hydroxide (NaOH) with different Na_2O concentrations of 4%, 6%, 8%, and 10% by slag weight. Mechanical properties comprising compressive strength, flexural strength, and tensile strength before and after exposure were measured. Thermogravimetric analysis (Thermogravimetric analysis (TGA) and Derivative thermogravimetric (DTG)), X-ray diffraction (XRD), scanning electron microscope (SEM), and energy-dispersive X-ray spectroscopy (EDS) were also used for strength alteration explanation. The results indicated that Na_2O concentration influence on strength variation of AAS mortar was observed clearly at temperature range from ambient temperature to 200 °C. The melting alteration of AAS mortar after exposed to 1200 °C was highly dependent on concentrations of Na_2O.

Keywords: alkali-activated slag; elevated temperatures; Na_2O concentration; residual strength; brittleness; melting

1. Introduction

Blast furnace slag is a by-product that is formed by rapidly cooling the slag liquid from the furnace in cast iron manufacture [1]. For a long time, blast furnace slag is known as a mineral admixture that can be used for partial replacement of Portland cement in blended cement or concrete [1]. In recent decades, non-Portland cement binder, named as alkali activated slag (AAS), which is synthesized by mixing blast furnace slag with alkali hydroxide, carbonate, or silicate attracted a great deal of attention of many scientists due to its high strength, durability, and low environmental impact [2,3]. The main product of AAS is a low crystalline hydrated calcium silicate, like a C-S-H gel type with a low CaO/SiO_2 ratio [4].

The activation process of slag is highly dependent on the physical-chemical properties of blast furnace slag, the nature and dosage of activator and the curing condition [5,6]. When compared with sodium carbonate or sodium hydroxide activator, slag activated by sodium silicate liquid possessed the highest compressive strength [5]. The activation products are predominantly composed of sodium and calcium aluminosilicate hydrates (C-A-S-H and N-A-S-H), as well as some hydrotalcite-like products when using sodium hydroxide or sodium silicate as activator [7–10]. Furthermore, urban and industrial glass waste was investigated to be used as a potential alkaline activator for blast furnace slag [11–13]. AAS binders was observed exhibit some advantages such as earlier and higher mechanical strengths, lower heat of hydration when compared with original Portland cement (OPC) and concretes [14]. However, slag mortar activated with sodium silicate liquid was reported to have a higher drying shrinkage and to be more brittle than ordinary Portland cement (OPC) mortar [15]. Previous studies indicated that none or little amount of Ca(OH)$_2$ was found in AAS system [16]. Consequently, AAS is expected to exhibit stronger resistance to extremely aggressive environments, such as chemical solution or high temperature exposure [17–21].

In recent years, many publications investigated the thermal behavior of various AAS material when exposed to elevated temperatures. Zuda, L. et al. [22–24] studied the alteration of sodium silicate powder activated slag mortar using quartz sand [22,23] and electrical porcelain [24] as fine aggregate subjected to high temperatures up to 1200 °C. The results presented a decrease in residual compressive strength of exposed mortar and obtained the lowest value of approximately 20% at 800 °C. However, the residual compressive strength greatly increased from 800 to 1200 °C due to sintering phenomena between binder matrix and aggregate. At 1200 °C, the remaining strength attained about 87% of unexposed mortar strength in the case of using quartz sand as fine aggregate and a doubling of original mortar strength when using electrical porcelain. Guerrieri, M. et al. [25] studied the effect of high temperatures up to 1200 °C on properties of AAS concrete activated by powdered sodium metasilicate and hydrated lime. The study showed that the residual strength of specimens was approximately 76%, 73%, 46% and 10% of unexposed specimen strength when exposed to 200 °C, 400 °C, 600 °C, and 800 °C, respectively. Moreover, the fire performance of AAS mortar cured in two different regimes (ambient and heat curing condition) with exposure temperatures from 200 to 800 °C was investigated by Türker, H.T. et al. [26]. The results illustrated that the strength of ambient curing specimens at 200 °C increased approximately 20% when compared to reference specimens, while a strength decrease was observed in heat curing mortar.

Many previous studies concluded that the nature of activator had significant effect on the alkali-activated slag properties, such as strength, microstructure, and shrinkage [5,15,27–30]. However, there are few publications in literature that have focused on the influence of activator on the AAS performance when subjected to elevated temperatures. Chi, M.C. [31] studied durability in high temperatures environment to 800 °C of AAS concrete while using alkaline activator with different concentrations of 4%, 5%, and 6% of Na$_2$O by slag weight. The results exhibited high temperatures resistance of concrete was improved when increasing the Na$_2$O concentration. Rashad, A.M. et al. [32] studied the effect of elevated temperatures on the AAS paste activated by Na$_2$SO$_4$ with concentrations 1%, 3% of Na$_2$O equivalent by slag mass. The sample compressive strength was observed to increase slightly with an increase Na$_2$O concentration after exposure to temperatures from 600 to 800 °C. Properties alteration of AAS paste with different sodium silicate concentration of 3.5%, 5.5%, 6.5%, 10.5% Na$_2$O by slag weight when exposed to high temperatures up to 1000 °C was also investigated by Rashad, A.M. et al. [33]. The study indicated that the paste strength before and after exposure increased as the concentration of Na$_2$O increased. Nevertheless, until now there has not been any publication which studied the influence of activator with different alkaline concentrations on the behavior of AAS mortar upon exposure to temperatures ranging from 200 to 1200 °C. Furthermore, the tensile strength of AAS mortar that was exposed to high temperatures has not also been investigated in previous studies. Consequently, the main goal of this paper is to determine the various Na$_2$O concentration alkaline activated slag mortar mechanical strength and microstructure alteration after exposure to temperatures up to 1200 °C. Mechanical strength of mortar comprise compressive strength, flexural strength, and tensile strength.

2. Materials and Methods

2.1. Material Characterization

Blast furnace slag, which originated from South Korea, was used to synthesize the alkali activated slag mortar in current research. Slag has a specific surface area of 435 m^2/kg (Blaine) and a density of 2.9 g/cm^3. The activity index of slag at seven days and 28 days is 97% and 112%, respectively. The chemical composition and XRD analysis results of used slag material are presented in Table 1 and Figure 1. The XRD analysis diffractogram displays a wide diffusive hump between 25° and 35°, indicating that slag is mostly amorphous.

Table 1. Chemical composition of used blast furnace slag.

Oxide	SiO$_2$	CaO	Al$_2$O$_3$	Fe$_2$O$_3$	MgO	SO$_3$	Na$_2$O	K$_2$O	LOI
(%)	33.81	41.24	15.19	0.41	5.54	2.51	0.25	0.61	0.18

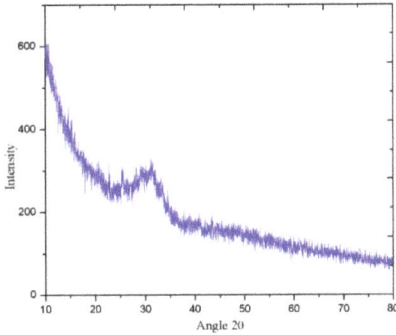

Figure 1. X-ray diffraction (XRD) pattern of blast furnace slag.

For material synthesis, blast furnace slag was activated by the alkaline activator, which was a combination of sodium silicate solution (water glass) and sodium hydroxide. Sodium silicate is liquid form with chemical composition comprising 26.4% Na$_2$O, 8.2%SiO$_2$, and 65.4% H$_2$O by mass. Sodium hydroxide (NaOH) pellets was dissolved in sodium silicate solution to decrease the silica modulus to 1 and Na$_2$O dosage of 4%, 6%, 8%, 10% by slag weight. The alkaline activator was prepared prior to mixing with slag 24 h. Local natural river sand (silica sand) with nominal maximum size of 4 mm and fineness modulus of 2.45 was used as fine aggregate to make mortar samples.

2.2. Mixture Proportion

The fine aggregate to slag mass ratio was 2.75. The alkaline activator portion was determined by dosage of Na$_2$O per cent by slag weight. Four mortar mixtures with different activator concentration of 4%, 6%, 8%, and 10% Na$_2$O by slag weight were named as A4, A6, A8, and A10, respectively. The water amount was adjusted to attain a water to solid ratio of 0.45 for all mortar mixtures. The solid portion included slag and solid component in alkaline activator.

2.3. Method

Blast furnace slag and fine aggregate were initially mixed in a 5 L capacity planetary blending machine for 1 min. Then, the mixture of alkaline activator and diluted water was poured into the machine bowl and continuously mixed for 2 min, followed by resting time of 1 min. During the resting period, the unmixed solids were scrapped from the sides and paddle into the mixing bowl. The whole mixture was mixed once again for 1 min. For casting, the fresh mortar mixture was poured into moulds of different shapes, depending on each mechanical property tests; 50 mm cube triplicate moulds for compressive strength test, 40 × 40 × 160 mm^3 moulds for flexural strength test, and number "8" shaped moulds for tensile strength test. The specimens were then vibrated by using a vibrating table for 1 min to release any residual air bubbles. To prevent water evaporation, the specimens were covered with a thin plastic sheet and kept in the laboratory environment with a temperature of 20 ± 5 °C and humidity of 60 ± 5% for one day. The specimens were unmoulded and cured in the same condition for more 27 days prior to subject to high temperatures.

The mortar strengths were determined after curing period of 28 days according to standard ASTM C109 for compressive strength [34], ASTM C348 for flexural strength [35], and ASTM C190 for tensile strength [36]. The specimens, which were tested in the age of 28 curing days without exposure to elevated temperatures, were called the reference specimens or unexposed specimens. After 28 days of curing, the specimens were dried in oven at temperature of $105 \pm 1\ °C$ for 24 h prior to subject to high temperatures of 200 °C, 400 °C, 600 °C, 800 °C, 1000 °C, and 1200 °C (T_c) by using an electrical heated furnace. For each temperature, the specimens were placed in the furnace and then heated at a rate of approximately 6.67 °C/min to obtain the determined temperature (T_c). When reaching the target temperature (T_c), the furnace temperature was maintained for 2 h. After that, the specimens were left in the furnace to cool naturally to ambient temperature. The furnace temperature versus time schedule is presented in Figure 2.

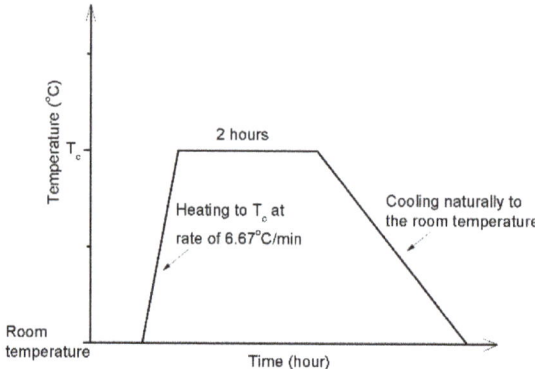

Figure 2. Temperature profile curve of furnace.

The exposed specimens were tested to determine the residual compressive strength, flexural strength, and tensile strength. After the compressive strength test, selected debris was immersed in acetone for three days to the stop hydration reaction. The debris was then filtered from acetone and dried in desiccator under vacuum. A part of dried samples was grounded and screened by using a 63 μm sieve. Fine particles passing a 63 μm sieve were used to analyze by X-ray diffraction (XRD) and thermogravimetric analysis (Thermogravimetric analysis (TGA)/Derivative thermogravimetric (DTG)) method. Nominated pieces were investigated by the scanning electron microscopy (SEM) with energy dispersive X-ray spectroscopy (EDS).

3. Results and Discussion

3.1. Compressive Strength

The behavior of mortar specimens with different Na_2O concentrations when being exposed to elevated temperatures was examined by determining the residual mortar strength alteration, which is illustrated in Figure 3a. Figure 3b presents the change of exposed specimen strength in comparison with that of reference specimens. It can be seen from Figure 3 that the increase of Na_2O concentration resulted in enhancement of the unexposed specimen compressive strength at 28 days. This result is consistent with previous studies [15,28]. With higher concentration of Na_2O, the higher pH value of solution accelerated chemical reaction of alkali activation of slag, causing material strength gain. Increasing Na_2O concentration from 4% to 6% led to great strength gain of approximately 61.2%. Nevertheless, the strength gain decreased gradually with further increasing concentration of Na_2O above 6%. For instance, the A10 specimen strength was higher than that of A8 specimens by approximately 4.4%, whilst the A8 specimen strength increased 14.4% of A6 specimen strength.

Parallel to the hydration process acceleration due to high alkaline activator, alkali activated slag exhibits high shrinkage deformation resulted from high sodium content [28,29] when the concentration of Na$_2$O increases. This may explain for reduction in strength gain of AAS mortar when increasing Na$_2$O concentration.

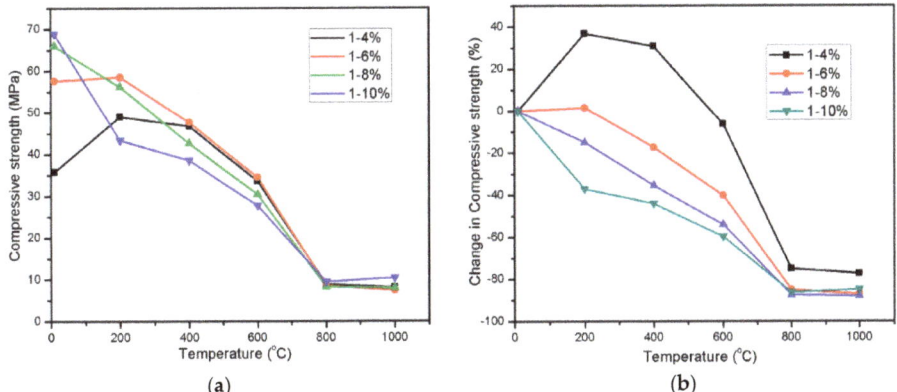

Figure 3. Residual compressive strength (**a**) and change in compressive strength (**b**) of alkali activated slag (AAS) mortar.

The residual compressive strengths of exposed specimens were highly dependent on Na$_2$O concentration especially with the exposure temperature range from 200 to 400 °C. At 200 °C, the A4 specimens had a remarkable strength increase of 36.9% in comparison with strength of unexposed specimens. Increasing the exposure temperature up to 400 °C led to reduction in strength gain, but the residual strength at 400 °C was still higher than reference strength of specimens by 30.7%. This result is similar to study of Türker, H.T. et al. [26] about the ambient cured mortar sample. According to Türker, H.T. et al. [26], this strength enhancement was likely caused by heating effect accelerated the hydration process. However, the strength gain of A4 mortar specimen exposed to 200 and 400 °C was observed to decrease or be absent with further increase of Na$_2$O concentration. Expose to 200 °C, the A6 specimens exhibited 1.6% enhancement of compressive strength whilst both A8 and A10 specimens possessed the great strength loss of 14.8% and 37%, respectively. It is noticeable that the loss of strength value of A10 mixture (37%) was equivalent to the strength gain value of A4 mixture (36.9%) at 200 °C. These results indicate that the enhancement in compressive strength at 200 °C decreases when increasing the concentration of Na$_2$O. Influence of heating treatment on strength development of alkali-activated slag material was investigated in many previous studies [27,37,38]. Elevated temperature curing greatly accelerates strength gain in sodium silicate-activated slag material at early-age [37,39]. Furthermore, extension the length of time the samples were kept in ambient condition prior to heating treatment was observed to be unbeneficial for strength gain due to heating curing [26,37]. On the other hand, Gebregziabiher, B.S. et al. [39] found that the sample with elevated temperature curing exhibited the higher enhancement in strength as sodium oxide (Na$_2$O) dosage increased. However, above results reveals that the strength gain due to later heating treatment occurs in sample that is activated by alkaline solution with low Na$_2$O dosage.

With further exposing temperatures beyond 400 °C, A4 specimens exhibited slight strength reduction of 6.5% at 600 °C and considerable degradation in strength of 75.1% at 800 °C. The strength of A6 specimen gradually decreased to 17.3% and 40.2% at exposure temperature of 400 °C and 600 °C, respectively. The strength loss of both A8 and A10 specimens gradually alleviated when increasing temperature from 200 to 400 °C and 600 °C. It was noticeable from Figure 3 that all of the mortar mixtures lost significant strength and those strength values converged at the temperature of 800 °C.

The slope of strength reduction line at temperature range from 400 to 600 and from 600 to 800 °C was observed to decrease when increasing the Na$_2$O concentration from 4% to 10%. Throughout elevated temperature range from 200 to 600 °C, the A6 specimens exhibited the highest strength value among four mixtures specimens, whilst the lowest strength was observed in the mortar mixture with Na$_2$O concentration of 10%.

There was a minor change in compressive strength of all the mixtures at temperature range 800–1000 °C. In comparison with residual strength at 800 °C, A4, A6, and A8 specimen at 1000 °C exhibited the slight reduction in strength of 9%, 14.6%, and 3.6%, respectively, whilst A10 attained strength gain of 9.5%. This slight strength gain at temperature range 800–1000 °C of AAS mortar is similar to previous research [23,24], in which slag was activated by dried sodium silicate. With the highest Na$_2$O concentration as 10%, the A10 specimens possessed the highest compressive strength at 800 and 1000 °C. It is obvious that the A4 specimens with the lowest Na$_2$O concentration 4% attained the greatest strength increase and the lowest strength deterioration upon heating to high temperatures. During heating examination up to 1000 °C, there was no sign of spalling for all of the mixtures mortar, but a great amount of small cracks appeared on specimen surface at 800 °C and 1000 °C (Figure 4).

Figure 4. Photographs of AAS mortar samples activated with different concentration of Na$_2$O before and after exposure to high temperatures from 200 to 1000 °C.

Raising the examination temperature up to 1200 °C resulted in deformation of all the mixture specimens due to melting phenomena. Therefore, the mechanical strength values could not be evaluated for samples after being exposed to 1200 °C. It is noticeable from Figure 5 that the deformed appearance was observed to be less in the specimens with higher concentration of Na$_2$O.

Figure 5. Photograph of AAS mortar samples after exposure to 1200 °C.

According to previous studies [15,40,41], the brittleness of a material is evaluated by the ratio of flexural strength to compressive strength and the angle of internal friction. Angle of internal friction from Mohr envelope can be shown as $\varphi = 0.5 \times \text{ARCTAN}((CS - TS)/\text{SQRT}(CS \times TS))$, where CS and TS are the compressive and tensile strength. The brittleness of a material increases when ratio of flexural strength to compressive strength decreases and the angle of internal friction increases [40,41]. It can be seen from result in Table 2 that raising the Na_2O concentration led to an increase in the brittleness property of AAS mortar in this study.

Table 2. Mechanical strength, ratio of flexural strength to compressive strength (FS/CS) and internal friction angle (φ) of AAS mortar at 28 days.

Mixture	Compressive Strength CS (MPa)	Flexural Strength FS (MPa)	Tensile Strength TS (MPa)	FS/CS	φ (Radian)
A4	35.8	6.8	1.54	0.18994	0.67868
A6	57.7	8.9	2.2	0.15425	0.68526
A8	66	9.3	1.64	0.14091	0.70527
A10	68.9	8.7	1.67	0.12627	0.70629

3.2. Flexural Strength

Flexural strength of AAS mortar in ambient temperature was highly dependent on Na_2O concentration of activator. Increasing the Na_2O concentration from 4% to 6% and 8% led to mortar strength gain of 30.9% and 36.8%, respectively. This result is consistent with finding of Duran Atiş, C. et al. [15]. However, the strength gain decreased slightly to 27.9% with a further increase in Na_2O concentration to 10%. Flexural strength is more sensitive to micro-cracks than compressive strength [42]. The AAS mortar with higher Na_2O concentration was observed to be more brittle, resulting in more micro-cracks due to higher shrinkage deformation [15]. Raising the concentration of Na_2O led to not only the reaction acceleration but also the higher shrinkage deformation. This could explain the reason why there was the flexural strength reduction in mortar with Na_2O concentration of 10%.

Exposing mortars to 200 °C diminished the flexural strength of mortar significantly with all Na_2O concentrations. More loss in strength was observed in samples with Na_2O concentration of 4 and 10% than that of 6 and 8% Na_2O samples. This flexural strength reduction could be attributed to further micro-crack formation resulted from shrinkage. As exposing to high temperature of 200 °C, ambient cured AAS mortar specimen experienced shrinkage which was combination of drying shrinkage due to evaporation of water from specimen and chemical shrinkage. According to Gu, Y.-M. et al. [42], the rapid reaction at relatively higher temperatures resulted in larger chemical shrinkage and induced micro-cracks in matrix at early ages, which developed with aging although the hardened paste got more compact. Increasing the exposing temperature from 200 to 400 °C and 600 °C alleviated the rate of reduction in flexural strength greatly. It is noticeable from Figure 6 that A4 specimen with the lowest Na_2O concentration exhibited slight strength gain at exposing temperature range from 400 to 600 °C. The flexural strength deterioration increased when exposed to temperature from 600 to 800 °C. As seen in Figure 6, flexural strength behavior of AAS mortar in temperature range from 800 to 1000 °C was similar to that of mortar compressive strength. In comparison with residual strength of mortar at 800 °C, the A4, A6, and A8 specimens at 1000 °C exhibited no change or slight reduction in strength, whilst the slight strength increase occurred in A10 specimen. Throughout the exposed temperatures, the residual strength of A6 and A8 specimens was equivalent each other. The flexural strength variation trend of mortar was different to that in result of some previous studies [23,24]. According to Zuda, L. et al. [23], there was no chance in flexural strength of AAS mortar, which used quartz sand as fine aggregate when exposing to 200 °C. When using electrical porcelain as fine aggregate of AAS mortar, Zuda, L. et al. [24] found that mortar flexural strength altered slightly at temperature range from 200 to 1000 °C.

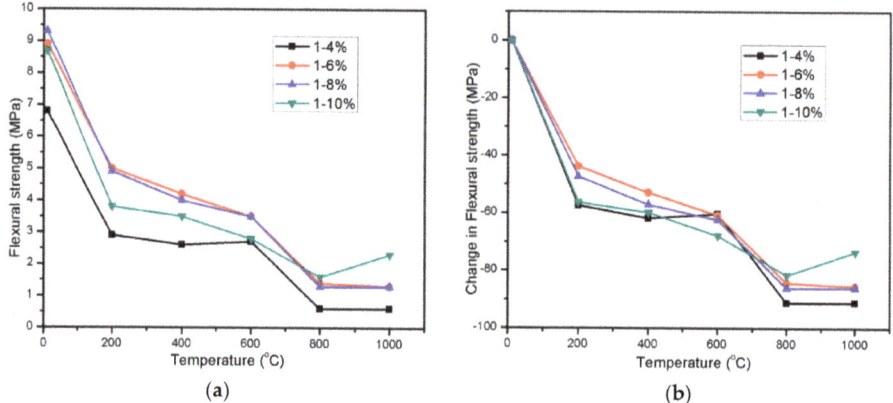

Figure 6. Residual flexural strength (**a**) and change in flexural strength (**b**) of AAS mortar.

3.3. Tensile Strength

The tensile strength which was determined by testing "8" shaped mortar specimens at ambient temperature and after exposure to high temperatures is given in Figure 7a. Figure 7b presents the relative tensile strength of mortar that is exposed to high temperatures in comparison with unexposed mortar strength. Similar to compressive and flexural strength, the Na_2O concentration of activator has an important role in tensile strength of unexposed specimens. The AAS mortar using activator with Na_2O concentration of 6% possessed the highest tensile strength value, which was approximately 1.43, 1.34, and 1.32 times higher than that of A4, A8, and A10 specimen, respectively. Raising Na_2O concentration in activator to 6% led to grow of the tensile strength value, however the strength was investigated to decrease with further increase of Na_2O concentration beyond 6%. According to Duran Atiş, C. et al. [15], there is direct correlation between the tensile strength and the brittleness of AAS mortar. The mortar with higher brittleness is higher shrinkage, and thus, the tensile strength is lower. The mortar is more brittle, the tensile strength is lower due to cracking by shrinkage deformation resulted from high Na concentration. When exposed to high temperatures from 200 to 800 °C, the tensile strength of all mixture AAS mortar decreased rapidly and then converged at temperature of 800 °C. The tensile strength of AAS mortar is strongly dependent on the paste-aggregate bond strength [43]. As exposing to high temperatures, the thermal incompatibility between AAS paste and fine aggregate resulted in weakening the paste-aggregate bond. In addition to weak AAS paste-aggregate bond, the development of pre-existing micro-cracks as well as new crack formation due to thermal shrinkage also caused deterioration in tensile strength of mortar. Throughout high temperature range from 200 to 600 °C, the A6 mortar specimen still possessed the highest residual strength in 4 mortar mixtures.

It can be seen in Figure 7 that Na_2O concentration has apparent influence on residual tensile strength of mortar at temperature of 1000 °C. In comparison with strength of mortar at 800 °C, the strength of A6, A8, and A10 mortar specimens at 1000 °C was observed to increase slightly, whilst strength loss occurred in the A4 mortar. The AAS mortar with higher Na_2O concentration exhibited higher tensile strength enhancement. For instant, the residual strength of A10 specimen after exposed to 1000 °C was 2.3 times higher than that of mortar at 800 °C. This ratio decreased to 1.8, 1.2, and 0.6 for mortar with Na_2O concentration of 8%, 6%, and 4%, respectively.

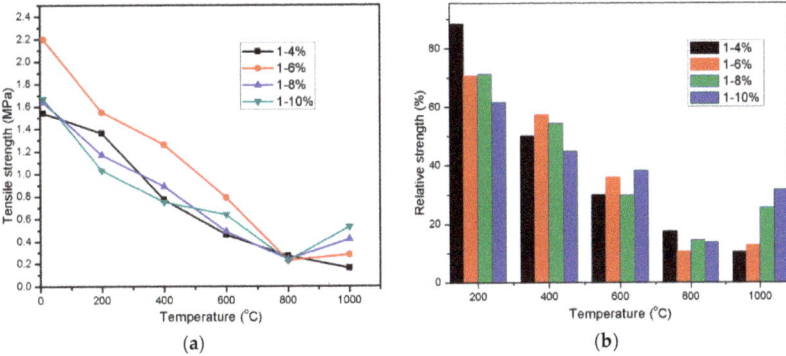

Figure 7. Residual tensile strength (**a**) and relative tensile strength (**b**) of AAS mortar.

According to above results, there was clear correlation between the brittleness property and residual mechanical strength of AAS mortar exposed to high temperature range from laboratory temperature to 200 °C and from 800 to 1000 °C. Relation between Na$_2$O concentration and the mechanical strength variation in temperature range from ambient to 200 °C and from 800 to 1000 °C is, respectively, shown in Figure 8a,b, respectively. After exposed to 200 °C, the loss in compressive strength was higher in AAS mortar with higher brittleness. In contrast, the mechanical strength of the AAS mortar had a tendency to increase with increasing the brittleness at temperature range 800–1000 °C.

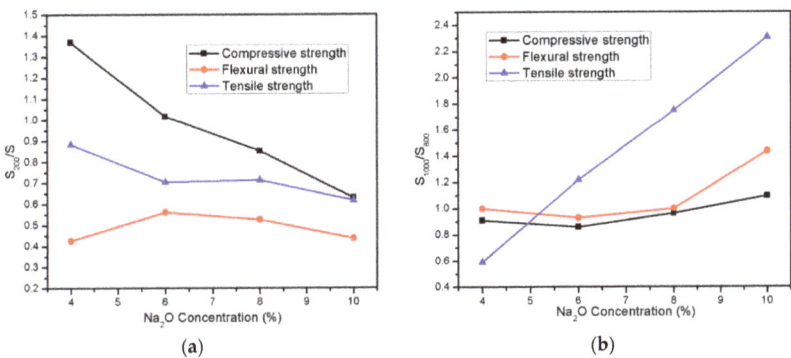

Figure 8. Relation between Na$_2$O concentration and the strength of AAS mortar after exposure to high temperature range: (**a**) From laboratory temperature to 200 °C; and, (**b**) From 800 to 1000 °C.

3.4. Thermogravimetric Analysis

Figure 9 presents the result of Thermogravimetric analysis (TGA) and Derivative thermogravimetric (DTG) of mortar with different Na$_2$O concentration of 4%, 6%, 8%, and 10%. For four samples, the main DTG peak centered at approximately 90 °C is result of dehydration of calcium silicate hydrate (C-S-H) [44]. It is noted from Figure 9b that the DTG peak of C-S-H is larger and sharper when increasing the Na$_2$O concentration in mortar. It reveals that more hydration product C-S-H is formed in AAS mortar with a higher concentration of Na$_2$O. This is consistent with the sharp mass loss of A10 and A8 mortar sample due to release of free water and OH groups from matrix [26] before around 120 °C (TGA curve in Figure 9a). The rapid migration and evaporation of water resulted in more micro-cracks in the mortar structure. This also may explained the reason why the significant deterioration in compressive strength occurred in A8 and A10 mortar when exposed to 200 °C. The reduction in mass loss was observed in all mixture specimens for temperature higher 150 °C. On the other hand, the wide DTG peak at approximately 550 °C due to decomposition of calcite ($CaCO_3$) [45] was apparently observed to be present in A4, A6, and A8 samples. These DTG peaks are consistent with sharp reduction in mass of A4, A6, and A8 sample in temperature range from 450 to 560 °C in TGA curve. Furthermore, the slope of mass reduction line in this temperature range increased with an increase of Na_2O concentration. The significant degradation in mass reduction occurred abruptly at approximately temperature of 560 °C. Beyond this temperature, there was minor change in the percentage of residual mass and all of the samples possessed the minimum mass value at 1000 °C. It can be seen from Figure 9a that the elevated temperatures treatment caused the highest mass loss in mortar with Na_2O concentration of 8%, followed with Na_2O concentration of 10%, 6%, and 4%, respectively.

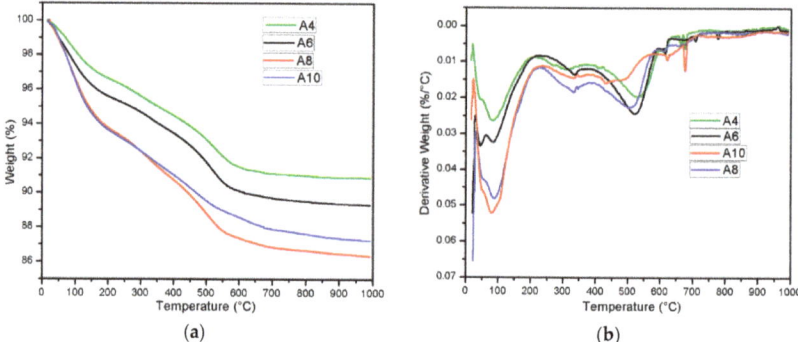

Figure 9. Thermogravimetric analysis (TGA) analysis (**a**) and Derivative thermogravimetric (DTG) analysis result (**b**) for AAS mortar at 28 days.

3.5. XRD Analysis

Figure 10 displays the XRD traces of unexposed AAS mortar with different Na_2O concentrations after 28 days of curing in ambient condition. The predominant crystalline phase of Quartz (SiO_2) was detected with minor reflections of Albite ($NaAlSi_3O_8$), Calcite ($CaCO_3$), and calcium silicate hydrate ($CaO \cdot SiO_2 \cdot nH_2O$) (C-S-H). From previous studies [33,45], the peak of C-S-H was present in overlapping with trace of Calcite ($CaCO_3$). The XRD trace of Calcite ($CaCO_3$) was identified clearly in A4, A6, and A8 mortar samples. However, Calcite reflection was not found in mortar specimen with Na_2O concentration of 10%. This XRD result is consistent with DTA and TGA analysis.

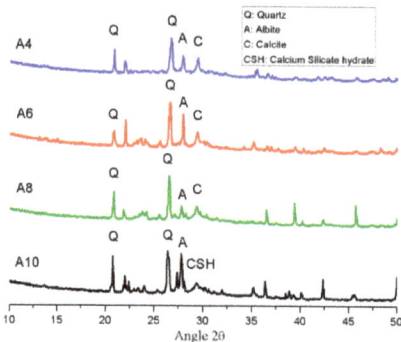

Figure 10. XRD patterns of AAS mortar with different concentration of Na_2O before exposure.

3.6. Microstructural Analysis

The SEM micrograph in Figure 11 shows the microstructure of AAS mortar activated by alkaline solution with different Na_2O concentrations at the age of 28 days curing in ambient condition. It is noted from Figure 11 that the AAS mortar is highly dependent on concentration of Na_2O. The sample with concentration of 4% Na_2O exhibited a porous microstructure and unhydrated slag particles. Figure 12a presents the EDS trace of spot marked 1 on A4 specimen surface (Figure 11) where the dominant elements were detected to be Si and Al, whilst Ca element was present with minor amount. The Calcium amount is not enough to make a reaction with Silicate component for C-S-H formation. These results prove that the low activation process of slag was resulted from low alkaline activator with Na_2O concentration of 4%. As seen in Figure 11b–d, the structure became denser and well-packed with increasing the concentration of Na_2O. Specimens A6, A8, and A10 display micro-cracks due to high shrinkage deformation. On the other hand, the EDS spectra in Figure 12b–d reveals that Ca element amount at spots marked 2, 3, 4 in Figure 11b–d increased significantly to be dominant with Si element when increasing concentration of Na_2O. Furthermore, EDS spectra (Figure 12e) of spot 5 at Figure 11d indicates that there was free sodium component in A10 mortar sample. Furthermore, Sodium substituted calcium silicate hydrate (N-C-S-H) with low Ca/Si is known as the main reaction product of alkali activated slag [46,47]. Malolepszy, J. [48] postulated the formation of a solid solution of Na_2O-CaO-SiO_2-H_2O (N-C-S-H), since Na^+ ions in alkali-activated cement have a very low solubility in water. It is well known that the Ca/Si ratio has a significant influence on properties of C-S-H. EDS image in Figure 12 reveals that Ca/Si ratio decreased with raising Na_2O concentration from 6% to 10%. The Ca/Si ratio reduction causes an improvement in binding ability of C-S-H [46,49,50], led to a compressive strength gain of AAS mortar.

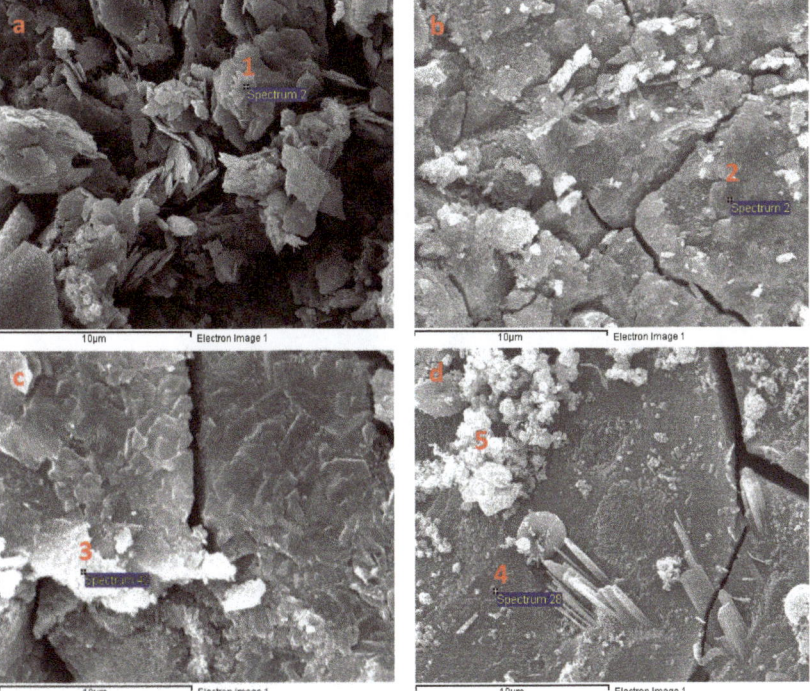

Figure 11. Scanning electron microscopy (SEM) micrograph of fracture surface of AAS mortar at 28 days: (**a**) A4; (**b**) A6; (**c**) A8; and, (**d**) A10.

The microstructure transformation of AAS mortar after exposure to elevated temperatures up to 1200 °C is given in Figures 13–15. It is noticeable that, as samples exposed to 200 °C, the porous structure of unexposed mortar with the lowest Na_2O concentration of 4% transformed to be compact structure with few hair-line micro-cracks. This change could be attributed to more hydration product resulted from the reaction acceleration due to the heating effect. Furthermore, migration of water before 200 °C occurred more easily from the porous structure of A4 mortar. This explains for the compressive strength gain of A4 mixture mortar at 200 °C. Moreover, Ca element amount at A4 fragment surface was investigated to increase highly from EDS spectra (Figure 16a) of spot marked 6 in Figure 13, revealing the formation of C-S-H. As seen in Figure 14, there was no significant change in microstructure of A6 mixture mortar at 200 °C. Contrary to A4 sample, the rapid migration and escape of a large amount of water from unexposed dense structure of A8 and A10 specimens caused the significant degradation in strength after exposure to 200 °C. This explains the reason why the microstructure of mortar with Na_2O concentration 10% possessed more cracks and rough surface at 200 °C (Figure 15). Micro-cracks were observed to be present on mortar surface (Figures 13–15) led to the drop of flexural and tensile strength when exposed to high temperatures.

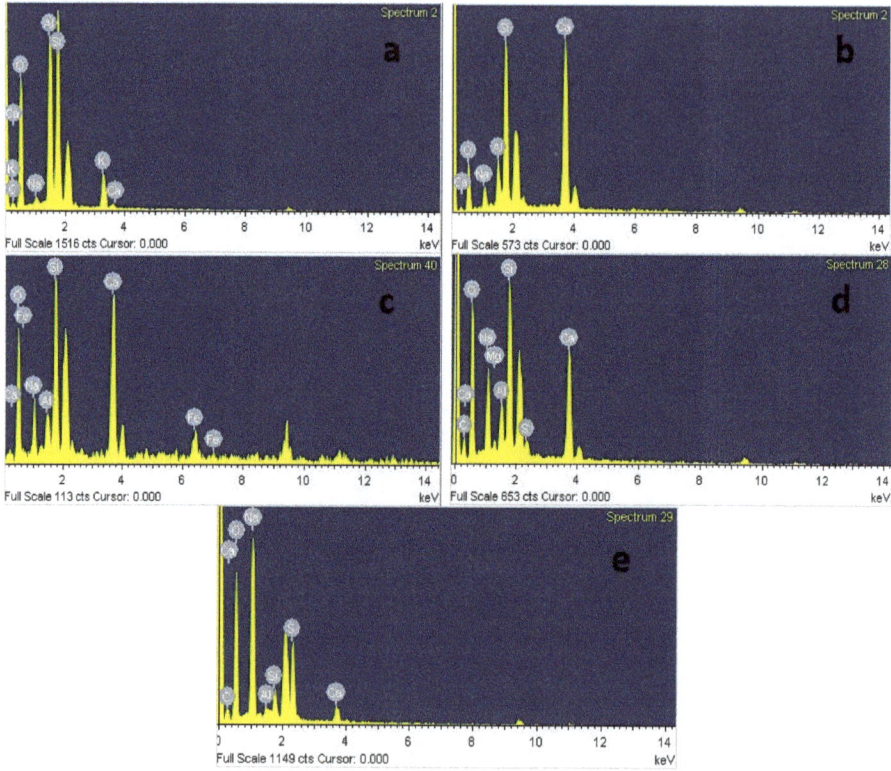

Figure 12. Energy dispersive X-ray spectroscopy (EDS) image of spots in Figure 11: (**a**) Spot labeled 1; (**b**) Spot labeled 2; (**c**) Spot labeled 3; (**d**) Spot labeled 4; and, (**e**) Spot labeled 5.

One of the main reasons for the deterioration in strength of AAS mortar at high temperature was thermal incompatibility between AAS matrix and fine aggregate. For exposure to elevated temperature, the AAS paste contraction occurred by the loss of water, whilst fine aggregate expands, resulting in weakening the bond between paste and aggregate. In addition, the occurrence of calcite decomposition process in mortars with Na$_2$O concentration of 4%, 6%, and 8% at a temperature of around 550 °C weakened mortar structure. This likely to explain reason why A10 specimen exhibited the lower deterioration rate in strength at temperature range from 400 to 600 °C. Furthermore, the great strength reduction result of AAS mortar from 600 to 800 °C is consistent with the significant damaged structure of mortar, which is presented in Figures 13–15.

Figure 13. SEM micrographs of fracture surface of A4 at different temperatures: (**a**) 200 °C; (**b**) 600 °C; (**c**) 800 °C; (**d**) 1000 °C; and, (**e**) 1200 °C.

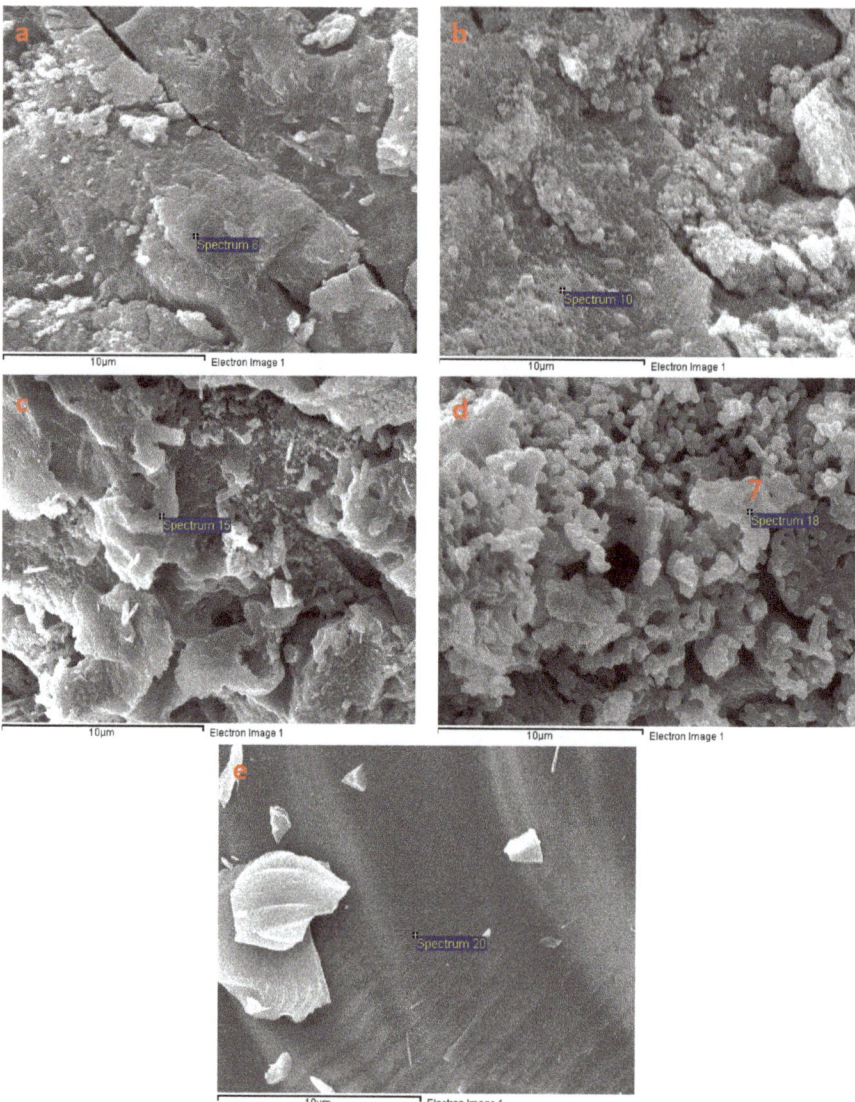

Figure 14. SEM micrographs of fracture surface of A6 at different temperatures: (**a**) 200 °C; (**b**) 600 °C; (**c**) 800 °C; (**d**) 1000 °C; and, (**e**) 1200 °C.

Increasing the exposing temperature from 800 to 1000 °C resulted in the significant alteration of AAS mortar structure due to sintering process, a significant increase in porosity and no sign of cracks was observed. This remarkable microstructure alteration may be the cause for slight change in mechanical strength of mortar from 800 to 1000 °C. As seen in Figures 13–15, alkali activated mortar structure at 1000 °C was denser with an increasing concentration of Na_2O. Figure 16 presents the EDS spectra of spots marked 7 in Figures 13–15, revealing a reduction in Ca/Si ratio when raising Na_2O concentration from 4% to 10%. The reduction of Ca/Si ratio in C-S-H structure indicates an enhancement in binding ability of C-S-H [49,50]. Both SEM and EDS result could explain for the higher strength of A10 mortar sample when exposing to 1000 °C. As sample exposed to temperature of 1200 °C, previous porous microstructure of all mortar mixtures was transformed to a considerable dense structure with smooth surface. The cracks and pores in structure at previous temperature were healed and filled by melting of AAS when exposed to 1200 °C. In spite of having highly dense structure, AAS mortar specimens observed to be deformed due to melting at 1200 °C and the intensity of deformation was higher in mortar with lower Na_2O concentration. This result could be attributed to binding ability of C-S-H in mortar structure, which was improved when increasing concentration of Na_2O.

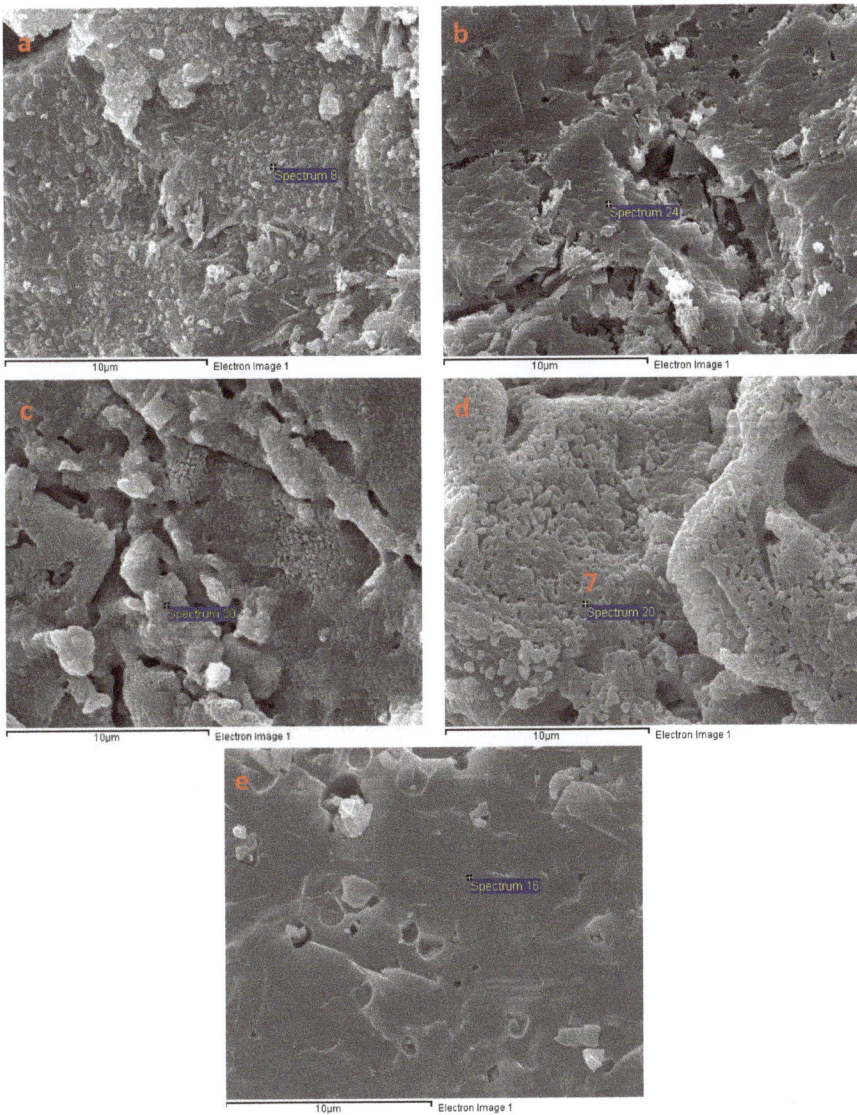

Figure 15. SEM micrographs of fracture surface of A10 at different temperatures: (**a**) 200 °C; (**b**) 600 °C; (**c**) 800 °C; (**d**) 1000 °C; and, (**e**) 1200 °C.

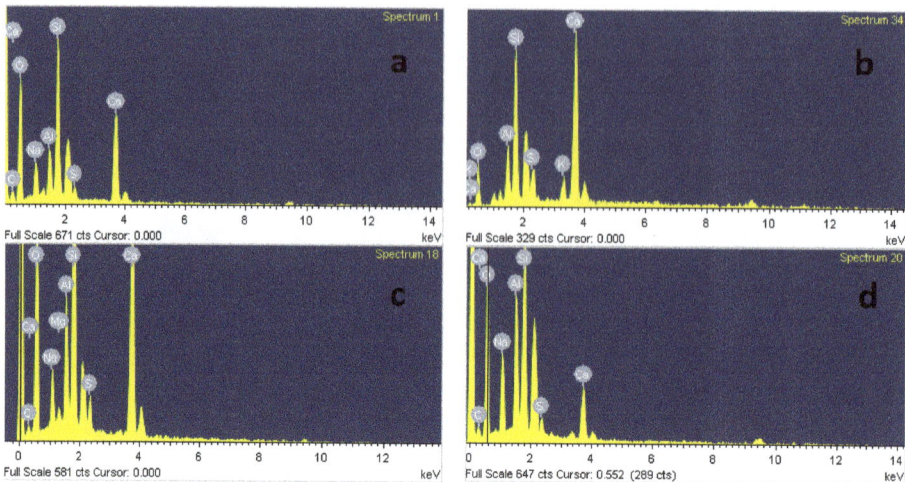

Figure 16. EDS image of: (**a**) Spot labeled 6 in Figure 13; (**b**) Spot labeled 7 in Figure 13; (**c**) Spot labeled 7 in Figure 14; and, (**d**) Spot labeled 7 in Figure 15.

4. Conclusions

Based on above experimental results and discussion, the study reveals the following conclusions:

The Na_2O concentration of alkaline activator has a great influence on mechanical strength of unexposed alkali-activated slag mortar, higher compressive strength with higher concentration of Na_2O. Moreover, raising the concentration of Na_2O led to increasing the brittleness of AAS mortar.

The compressive strength gain at 200 °C was observed in AAS mortar with a low Na_2O concentration of 4 and 6%, whilst mortar activated with Na_2O concentration of 8 and 10% exhibited a great reduction in strength.

The variation in residual flexural strength of AAS mortar was similar to tensile strength when exposed to high temperatures below 800 °C. The AAS mortar with Na_2O concentration of 6% exhibited the highest mechanical strength with exposing temperature below 800 °C.

The difference in residual mechanical strength of all mixture mortar was negligible at the exposure temperature range from 800 to 1000 °C. The mechanical strength of mortar at 1000 °C had a tendency to increase with higher Na_2O concentration and brittleness.

The highest mass loss after exposure was observed in alkali activated mortar with Na_2O concentration of 8%. Calcite ($CaCO_3$) was not found in alkali activated slag mortar with Na_2O concentration of 10%.

All mixture AAS mortar deformed significantly due to melting phenomena after being exposed to 1200 °C. The deformation was observed to be less with increasing concentration of Na_2O from 4% to 10%. The microstructure of mortar was changed to be highly dense with smooth surface when exposed to 1200 °C irrespective of Na_2O concentration.

Author Contributions: Both authors conceived and designed the experiments; T.T.T. performed the experiments, analyzed the results and wrote the original draft; H.-M.K. reviewed and edited the manuscript.

Funding: The research was supported by a grant (14-CRTI-B063773-03) from Infrastructure and transportation technology promotion research Program funded by Ministry of Land, Infrastructure and Transport of Korean government.

Conflicts of Interest: The authors declare no conflicts of interest.

References

1. Mehta, P.K.; Monteiro, P.J.M. *Concrete: Microstructure, Properties, and Materials*, 3rd ed.; McGraw-Hill: New York, NY, USA, 2006; p. 305, ISBN 0071589198.
2. OA, P. The action of alkalis on blast furnace slag. *J. Soc. Chem. Ind.* **1940**, *59*, 191–202.
3. Rovnaník, P.; Bayer, P.; Rovnaníková, P. Characterization of alkali activated slag paste after exposure to high temperatures. *Constr. Build. Mater.* **2013**, *47*, 1479–1487. [CrossRef]
4. Pacheco-Torgal, F.; Castro-Gomes, J.; Jalali, S. Alkali-activated binders: A review. Part 1. Historical background, terminology, reaction mechanisms and hydration products. *Constr. Build. Mater.* **2008**, *22*, 1305–1314. [CrossRef]
5. Bakharev, T.; Sanjayan, J.G.; Cheng, Y.B. Alkali activation of Australian slag cements. *Cem. Concr. Res.* **1999**, *29*, 113–120. [CrossRef]
6. Bernal, S.; De Gutierrez, R.; Delvasto, S.; Rodriguez, E. Performance of an alkali-activated slag concrete reinforced with steel fibers. *Constr. Build. Mater.* **2010**, *24*, 208–214. [CrossRef]
7. Wang, S.D.; Scrivener, K.L. Hydration products of alkali activated slag cement. *Cem. Concr. Res.* **1995**, *25*, 561–571. [CrossRef]
8. Brough, A.R.; Atkinson, A. Sodium silicate-based, alkali-activated slag mortars—Part I. Strength, hydration and microstructure. *Cem. Concr. Res.* **2002**, *32*, 865–879. [CrossRef]
9. Puertas, F.; Palacios, M.; Manzano, H.; Dolado, J.S.; Rico, A.; Rodríguez, J. A model for the C-A-S-H gel formed in alkali-activated slag cements. *J. Eur. Ceram. Soc.* **2011**, *31*, 2043–2056. [CrossRef]
10. Deir, E.; Gebregziabiher, B.S.; Peethamparan, S. Influence of starting material on the early age hydration kinetics, microstructure and composition of binding gel in alkali activated binder systems. *Cem. Concr. Compos.* **2014**, *48*, 108–117. [CrossRef]
11. Torres-Carrasco, M.; Tognonvi, M.T.; Tagnit-Hamou, A.; Puertas, F. Durability of alkali-activated slag concretes prepared using waste glass as alternative activator. *ACI Mater. J.* **2015**, *112*, 791–800. [CrossRef]
12. Puertas, F.; Torres-Carrasco, M. Use of glass waste as an activator in the preparation of alkali-activated slag. Mechanical strength and paste characterisation. *Cem. Concr. Res.* **2014**, *57*, 95–104. [CrossRef]
13. Torres-Carrasco, M.; Puertas, F. Waste glass as a precursor in alkaline activation: Chemical process and hydration products. *Constr. Build. Mater.* **2017**, *139*, 342–354. [CrossRef]
14. Puertas, F.; Amat, T.; Fernández-Jiménez, A.; Vázquez, T. Mechanical and durable behaviour of alkaline cement mortars reinforced with polypropylene fibres. *Cem. Concr. Res.* **2003**, *33*, 2031–2036. [CrossRef]
15. Duran Atiş, C.; Bilim, C.; Çelik, Ö.; Karahan, O. Influence of activator on the strength and drying shrinkage of alkali-activated slag mortar. *Constr. Build. Mater.* **2009**, *23*, 548–555. [CrossRef]
16. Jumppanen, U.-M.; Diederichs, U.; Hinrichsmeyer, K. *Material Properties of F-Concrete at High Temperatures*; Valtion Teknillinen Tutkimuskeskus: Espoo, Finland, 1986; p. 60.
17. Bakharev, T.; Sanjayan, J.G.; Cheng, Y.B. Resistance of alkali-activated slag concrete to acid attack. *Cem. Concr. Res.* **2003**, *33*, 1607–1611. [CrossRef]
18. Park, J.W.; Ann, K.Y.; Cho, C.G. Resistance of Alkali-Activated Slag Concrete to Chloride-Induced Corrosion. *Adv. Mater. Sci. Eng.* **2015**, *2015*. [CrossRef]
19. Bakharev, T.; Sanjayan, J.G.; Cheng, Y.B. Sulfate attack on alkali-activated slag concrete. *Cem. Concr. Res.* **2002**, *32*, 211–216. [CrossRef]
20. Rajamane, N.P.; Nataraja, M.C.; Lakshmanan, N.; Dattatreya, J.K.; Sabitha, D. Sulphuric acid resistant ecofriendly concrete from geopolymerisation of blast furnace slag. *Indian J. Eng. Mater. Sci.* **2012**, *19*, 357–367.
21. El-Didamony, H.; Amer, A.A.; Abd Ela-Ziz, H. Properties and durability of alkali-activated slag pastes immersed in sea water. *Ceram. Int.* **2012**, *38*, 3773–3780. [CrossRef]
22. Zuda, L.; Rovnaník, P.; Bayer, P.; Černý, R. Thermal properties of alkali-activated slag subjected to high temperatures. *J. Build. Phys.* **2007**, *30*, 337–350. [CrossRef]
23. Zuda, L.; Pavlík, Z.; Rovnaníková, P.; Bayer, P.; Černý, R. Properties of alkali activated aluminosilicate material after thermal load. *Int. J. Thermophys.* **2006**, *27*, 1250–1263. [CrossRef]
24. Zuda, L.; Rovnaník, P.; Bayer, P.; Černý, R. Effect of high temperatures on the properties of alkali activated aluminosilicate with electrical porcelain filler. *Int. J. Thermophys.* **2008**, *29*, 693–705. [CrossRef]

25. Guerrieri, M.; Sanjayan, J.; Collins, F. Residual compressive behavior of alkali-activated concrete exposed to elevated temperatures. *Fire Mater.* **2009**, *33*, 51–62. [CrossRef]
26. Türker, H.T.; Balçikanli, M.; Durmuş, I.H.; Özbay, E.; Erdemir, M. Microstructural alteration of alkali activated slag mortars depend on exposed high temperature level. *Constr. Build. Mater.* **2016**, *104*, 169–180. [CrossRef]
27. Altan, E.; Erdoğan, S.T. Alkali activation of a slag at ambient and elevated temperatures. *Cem. Concr. Compos.* **2012**, *34*, 131–139. [CrossRef]
28. Bilim, C.; Ati, C.D. Alkali activation of mortars containing different replacement levels of ground granulated blast furnace slag. *Constr. Build. Mater.* **2012**, *28*, 708–712. [CrossRef]
29. Krizan, D.; Zivanovic, B. Effects of dosage and modulus of water glass on early hydration of alkali-slag cements. *Cem. Concr. Res.* **2002**, *32*, 1181–1188. [CrossRef]
30. Chi, M.C.; Chang, J.J.; Huang, R. Strength and drying shrinkage of alkali-activated slag paste and mortar. *Adv. Civ. Eng.* **2012**, *2012*. [CrossRef]
31. Chi, M. Effects of dosage of alkali-activated solution and curing conditions on the properties and durability of alkali-activated slag concrete. *Constr. Build. Mater.* **2012**, *35*, 240–245. [CrossRef]
32. Rashad, A.M.; Bai, Y.; Basheer, P.A.M.; Collier, N.C.; Milestone, N.B. Chemical and mechanical stability of sodium sulfate activated slag after exposure to elevated temperature. *Cem. Concr. Res.* **2012**, *42*, 333–343. [CrossRef]
33. Rashad, A.M.; Zeedan, S.R.; Hassan, A.A. Influence of the activator concentration of sodium silicate on the thermal properties of alkali-activated slag pastes. *Constr. Build. Mater.* **2016**, *102*, 811–820. [CrossRef]
34. ASTM. ASTM C109-02 Standard Test Method for Compressive Strength of Hydraulic Cement Mortars. In *Annual Book of ASTM Standards*; American Society for Testing and Materials: West Conshohocken, PA, USA, 2002.
35. ASTM. ASTM C348-02 Standard Test Method for Flexural Strength of Hydraulic-Cement Mortars. In *Annual Book of ASTM Standards*; American Society for Testing and Materials: West Conshohocken, PA, USA, 2002.
36. ASTM. ASTM C190-85 Standard Test Method for Tensile Strength of Hydraulic Cement Mortars. In *Annual Book of ASTM Standards*; American Society for Testing and Materials: West Conshohocken, PA, USA, 1985.
37. Bakharev, T.; Sanjayan, J.G.; Cheng, Y.B. Effect of elevated temperature curing on properties of alkali-activated slag concrete. *Cem. Concr. Res.* **1999**, *29*, 1619–1625. [CrossRef]
38. Aydin, S.; Baradan, B. Mechanical and microstructural properties of heat cured alkali-activated slag mortars. *Mater. Des.* **2012**, *35*, 374–383. [CrossRef]
39. Gebregziabiher, B.S.; Thomas, R.J.; Peethamparan, S. Temperature and activator effect on early-age reaction kinetics of alkali-activated slag binders. *Constr. Build. Mater.* **2016**, *113*, 783–793. [CrossRef]
40. Kahraman, S.; Altindag, R. A brittleness index to estimate fracture toughness. *Int. J. Rock Mech. Min. Sci.* **2004**, *41*, 343–348. [CrossRef]
41. Gunsallus, K.L.; Kulhawy, F.H. A comparative evaluation of rock strength measures. *Int. J. Rock Mech. Min. Sci. Geomech. Abstr.* **1984**, *21*, 233–248. [CrossRef]
42. Gu, Y.-M.; Fang, Y.-H.; You, D.; Gong, Y.-F.; Zhu, C.-H. Properties and microstructure of alkali-activated slag cement cured at below- And about-normal temperature. *Constr. Build. Mater.* **2015**, *79*, 1–8. [CrossRef]
43. Wardhono, A.; Gunasekara, C.; Law, D.W.; Setunge, S. Comparison of long term performance between alkali activated slag and fly ash geopolymer concretes. *Constr. Build. Mater.* **2017**, *143*, 272–279. [CrossRef]
44. Rashad, A.M.; Zeedan, S.R.; Hassan, H.A. A preliminary study of autoclaved alkali-activated slag blended with quartz powder. *Constr. Build. Mater.* **2012**, *33*, 70–77. [CrossRef]
45. Rashad, A.M.; Sadek, D.M.; Hassan, H.A. An investigation on blast-furnace stag as fine aggregate in alkali-activated slag mortars subjected to elevated temperatures. *J. Clean. Prod.* **2016**, *112*, 1086–1096. [CrossRef]
46. Aydin, S.; Baradan, B. Effect of activator type and content on properties of alkali-activated slag mortars. *Compos. Part B Eng.* **2014**, *57*, 166–172. [CrossRef]
47. Aydin, S. A ternary optimisation of mineral additives of alkali activated cement mortars. *Constr. Build. Mater.* **2013**, *43*, 131–138. [CrossRef]

48. J, M. Some aspects of alkali activated cementitious materials setting and hardening. In Proceedings of the 3rd Beijing International Symposium on Cement and Concrete, Beijing, China, 27–30 October 1993; pp. 1043–1046.
49. Bakharev, T.; Sanjayan, J.G.; Cheng, Y.B. Effect of admixtures on properties of alkali-activated slag concrete. *Cem. Concr. Res.* **2000**, *30*, 1367–1374. [CrossRef]
50. Puertas, F.; Fernández-Jiménez, A.; Blanco-Varela, M.T. Pore solution in alkali-activated slag cement pastes. Relation to the composition and structure of calcium silicate hydrate. *Cem. Concr. Res.* **2004**, *34*, 139–148. [CrossRef]

 © 2018 by the authors. Licensee MDPI, Basel, Switzerland. This article is an open access article distributed under the terms and conditions of the Creative Commons Attribution (CC BY) license (http://creativecommons.org/licenses/by/4.0/).

Article

Double Feedback Control Method for Determining Early-Age Restrained Creep of Concrete Using a Temperature Stress Testing Machine

He Zhu, Qingbin Li *, Yu Hu * and Rui Ma

State Key Laboratory of Hydroscience and Engineering, Tsinghua University, Beijing 100084, China; zhuhe14@mails.tsinghua.edu.cn (H.Z.); marui14@mails.tsinghua.edu.cn (R.M.)
* Correspondence: qingbinli@tsinghua.edu.cn (Q.L.); yu-hu@tsinghua.edu.cn (Y.H.); Tel.: +86-010-62771015 (Q.L.); +86-010-62781161 (Y.H.)

Received: 1 May 2018; Accepted: 20 June 2018; Published: 25 June 2018

Abstract: Early-age restrained creep influences the cracking properties of concrete. However, conventional creep measurements require a large number of tests to predict the restrained creep as it is influenced by the combined effects of variable temperature, creep recovery, and varying compression and tension stresses. In this work, a double feedback control method for temperature stress testing was developed to measure the early-age restrained creep of concrete. The results demonstrate that the conventional single feedback control method neglects the effect of restrained elastic deformation, thus providing a larger-than-actual creep measurement. The tests found that the double feedback control method eliminates the influence of restrained elastic deformation. The creep results from the double feedback method match well with results from the single feedback method after compensation for the effects of restrained elastic deformation is accounted for. The difference in restrained creep between the single and double feedback methods is significant for concrete with a low modulus of elasticity but can be neglected in concrete with a high modulus of elasticity. The ratio between creep and free deformation was found to be 40–60% for low, moderate, and high strength concretes alike. The double feedback control method is therefore recommended for determining the restrained creep using a temperature stress testing machine.

Keywords: restraint; creep; double feedback method; concrete; temperature stress testing machine (TSTM)

1. Introduction

The cracking of massive concrete structures due to thermal stresses is a problem that has long been studied by engineers [1]. Dam concrete, a typical type of mass concrete, suffers varying temperature and strong restraint effects during the first days following casting. This influences early-age creep. The early-age creep of concrete can relax more than 50% of the restraint stress [2–4]. Creep is one of the most important properties that influence temperature stress in concrete due to relaxation effects [5].

In dam concrete, early-age creep has its own unique characteristics: The temperature in the concrete increases due to the hydration heat and must be controlled below a design value to avoid a large temperature gradient. The temperature stress is not a constant load, instead increasing and decreasing during the hydration process [6,7]. The early-age creep of concrete is also difficult to obtain because both the physical and chemical properties of concrete change simultaneously [4,8]. Therefore, the determination of early-age concrete creep under both varying temperature and restraint conditions is a challenging and a significant topic for research.

The influence factors on the creep, such as temperature, creep recovery, and loading at early-age have been widely studied. Temperature effects have been studied from the perspective of equivalent age [7], hydration degree [9], and transient creep [10,11], among others. Some calculation models, such as the B3 model [12], microprestress–solidification theory [13], rheological model [14,15], and degree of the hydration-based creep model [16,17] have been proposed to predict the effects of temperature on creep. Creep recovery is another important characteristic of creep. Rheological models, such as the modified Double Power Law [7], Kelvin-Voigt [18], and Maxwell [19] can effectively predict the recovery effect. Delsaute [20,21] combined a classical test and a repeated minute–scale–duration loading test to model the recovery effect.

Loading age can influence the creep magnitude, and the creep loading in early age is more significant than in later age. Existing creep models require modification to predict the extent of this early-age creep. For example, Østergaard [22] suggested a mathematical model for early-age creep by redefining a parameter in the B3 model. Similarly, Wei [23] redefined this parameter in the modified microprestress–solidification model to consider the effects of both temperature and early age on creep. Although the temperature, early age, and unloading factors have been extensively studied, the combined effect of these factors on the very early-age restrained creep of dam concrete is still difficult to accurately predict.

A temperature stress testing machine (TSTM) can be used to study early-age creep under the combined effects of restraint and temperature. The TSTM was developed by many researchers [3,24–28], and a multi-TSTM system controlled by a synchronous closed loop method was constructed by Zhu [29]. The restrained creep has been extracted from free and restrained specimens using a TSTM [25] based on the assumption of linear superposition, which was validated by [7,8,30,31]. In traditional TSTM tests controlled by the single feedback method, the restraint stress is also variable in a compensation cycle, and an additional restraint elastic deformation caused by the varied restraint stress is produced. This varied restraint stress has typically been ignored when decoupling the restrained creep [2–4,32].

In order to experimentally investigate the restrained creep of dam concrete under the effects of temperature and restraint conditions at early age, a double feedback control method was developed in this study based on the multi-TSTM system [29]. In the conventional creep test when applying a constant load, a series of tests to consider the varying temperature, alternating tension and compression stress, and different loading ages are required. Based on the multi-TSTM system, the restrained elastic deformation effect was then studied. Finally, a testing method for restrained creep using TSTM is proposed.

2. Materials and Methods

2.1. Double Feedback Control Method and Creep Calculation Method Using the TSTM

In a TSTM as shown in Figure 1a, the specimen deformation is measured directly by a deformation sensor embedded into the concrete and the load is monitored by the load cell. The TSTM can maintain the specimen in a restrained state by continuously checking the specimen deformation. When the preset deformation threshold is exceeded, the actuator of the TSTM is started to compensate for the strain, and the specimen is pushed or pulled to return it to its original length [25]. The number of the compensation cycle is indicated as 1, 2, ... i, and the time was marked as $t_1, t_2, \ldots t_i$.

In this work, a double feedback control method was developed. Both the load and the deformation of the specimen were monitored and automatically controlled by the computer in real-time. In addition to restoring the specimen to its original length when the deformation reached the preset threshold, the actuator was also working to maintain the load of the specimen as a constant value between the compensation cycles. As shown in Figure 2a, Case I represents the free specimen, and Case II represents the restrained specimen. The cumulative deformation curves are shown in Figure 2b, and the load history of the specimen controlled by double feedback method is shown in Figure 2c.

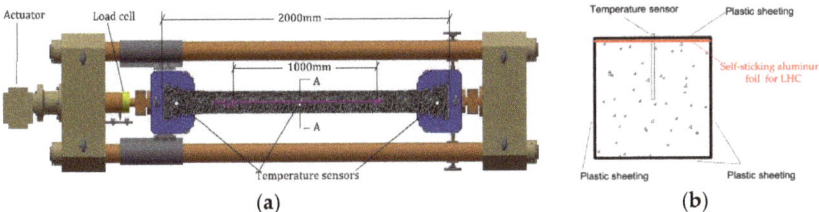

Figure 1. Schematic and the specimen geometry of a temperature stress testing machine (TSTM). (a) Schematic of TSTM; (b) A-A cross-section view.

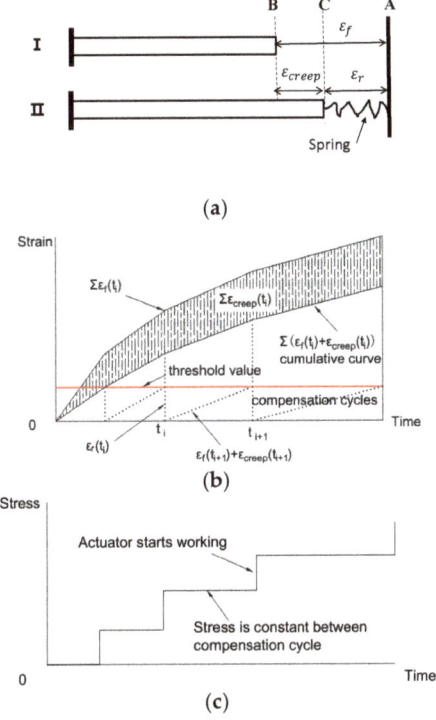

Figure 2. Restrained creep calculation method without considering the restrained elastic deformation, in which ε_r is the strain under the restrained condition, ε_f is the free deformation, and ε_{creep} is the restrained creep strain. (a) Deformation decomposition between the strain compensation cycles; (b) Cumulative curve of free deformation, elastic strain and restrained creep; (c) Stress history.

The deformation can be decomposed into restrained strain and creep as shown in Figure 2a,b, in which ε_f, ε_r, and ε_{creep} represent the free strain, restrained strain, and restrained creep, respectively. ε_r is the restrained strain recorded by the strain sensor, and the actuator starts to compensate when ε_r reaches the preset threshold, so the restrained strain is equal to the instantaneous elastic strain induced by the increment load applied by the actuator. The deformations conform to the geometric Equation (1), describing the relationship between the strain compensation cycles, so the restrained creep strain at time t_n can be calculated by Equation (2).

$$\varepsilon_{creep}(t_i) = \varepsilon_f(t_i) - \varepsilon_r(t_i) \tag{1}$$

$$\varepsilon_{creep}(t_n) = \sum_{i=1}^{n}\left[\varepsilon_f(t_i) - \varepsilon_r(t_i)\right] \tag{2}$$

The restrained creep in the TSTM controlled by the double feedback method can be calculated by Equation (2).

2.2. Singles Feedback Control Method and Creep Calculation Method Using the TSTM

For a traditional TSTM [4,25,33], the control method is called single feedback control method, in which only the deformation is checked to keep the specimen restrained. During the compensation cycle, the load applied to the specimen increases as the deformation increases to the threshold. Both the deformation and the load are variable during the compensation cycle as shown in Figure 3a,c. The difference between the single feedback method and proposed double feedback method is that the load is not a constant value between the strain compensations. The proposed double feedback TSTM is generally conducted as a creep relaxation combining test as both the restrained stress and strain are all variable [33]. An additional restrained elastic deformation, marked as ε_{re} in Figure 3a,b is produced due to the variable restrained stress during the compensation cycle.

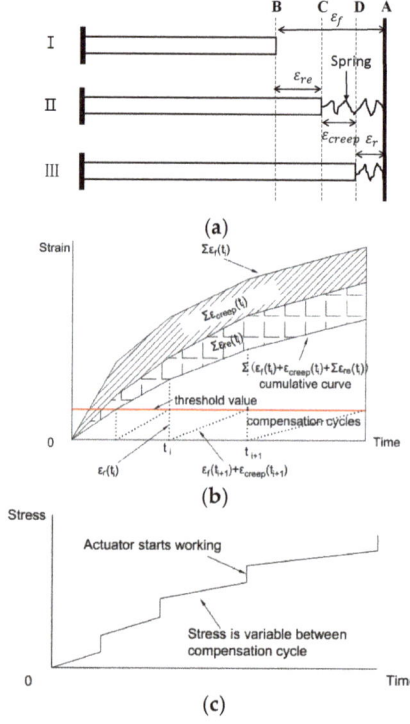

Figure 3. Restrained creep calculation considering the restrained elastic deformation, in which ε_{re} is the restrained strain. (**a**) Deformation decomposition between the strain compensation cycles of the single feedback control method; (**b**) Cumulative curve of free deformation, restrained elastic deformation, and restrained creep; (**c**) Stress history.

Under the single feedback method [25], the restrained elastic deformation has typically been ignored. The creep in a TSTM test controlled by the traditional single feedback method is also calculated by Equation (2), by decoupling the restrained creep without considering the restrained elastic deformation [2–4,32]. In reality, the restrained stress is accordingly variable during a compensation cycle, and an additional restrained elastic deformation will be produced. Consequently, the creep deformation calculated by Equation (2) will be influenced by the restraint stress.

In Figure 3a, Case I represents the free specimen, Case II represents the restrained specimen under the assumption that the concrete is elastic without considering the creep, and Case III represents the restrained specimen under which concrete is considered to be viscoelastic. Equation (3) can be derived from the force equilibrium equations as follows:

$$\Delta F_{re}(t_i) = E_{ce}(t_i) \times A_c \times \varepsilon_{re}(t_i) \tag{3}$$

where $\Delta F_{re}(t_i)$ is the restrained elastic load increment corresponding to the restored elastic deformation increment $\varepsilon_{re}(t_i)$ in compensation cycle i, and can be recorded by the load cell of the TSTM in a compensation cycle; A_c is the cross-sectional area of the concrete specimen; $E_{ce}(i)$ is the concrete elastic modulus, and $\varepsilon_{re}(t_i)$ is the restrained elastic deformation increment of the concrete under the restraint of the device during one adjustment cycle. Based on linear superposition, the deformation relationship is:

$$\varepsilon_{re}(t_i) = \Delta F_{re}(t_i) / [A_c \times E_{ce}(t_i)] \tag{4}$$

$$\varepsilon_f(t_i) = \varepsilon_{re}(t_i) + \varepsilon_{creep}(t_i) + \varepsilon_r(t_i) \tag{5}$$

Therefore, in a complete test period, the creep deformation at time t_n can be obtained as shown in Figure 3b and calculated by Equation (6) as follows:

$$\varepsilon_{creep}(t_n) = \sum_{i=1}^{n} \left\{ \varepsilon_f(t_i) - \varepsilon_r(t_i) - \Delta F_{re}(t_i) / [A_c \times E_{ce}(t_i)] \right\} \tag{6}$$

2.3. Materials

All the raw materials used for mixing the concrete were transported from the construction site of a super-high arch dam. The mix ratios of the concretes are listed in Table 1, in which the water-cement ratio was 0.50. Manufactured sand with a fineness modulus of 2.61 and apparent density of 2790 kg/m^3 was chosen as the fine aggregate. Limestone gravel with a diameter of 5–20 mm, an apparent density of 2790 kg/m^3, and a saturated surface dry water absorption rate of 0.21% by mass was employed as the coarse aggregate. The density of the fly ash was 2320 kg/m^3, and its fineness was 7.6%. The effect of the admixture on the concrete performance is given in Table 2. The constituents of the Portland cement and the fly ash are provided in Table 3. Two kinds of concrete, low-heat cement concrete (LHC) and moderate-heat cement concrete (MHC), were mixed in the same ratios given in Table 1. The only difference between the two mixes was the cement used, detailed in Table 3.

Table 1. Mix ratios of the concrete specimens used in the experiments (kg/m^3).

Water	Cement	Fly Ash	Sand	Gravel	Water-Reducing Admixture	Air-Entraining Admixture
130.00	169.00	91.00	727.58	1351.23	1.12	0.074

Table 2. Effect of the admixtures on the performance of the concrete.

Admixture	Mix Ratio (%)	Water Reduction Ratio (%)	Air Content (%)	Bleeding Rate (%)	Difference in Setting Time (min)	
					Initial Setting	Final Setting
Water reducing admixture	0.60	19.5	1.9	25	+260	+350
Air entraining admixture	0.008	6.5	5.0	35	+40	+70

Table 3. Chemical composition of the materials used (% mass).

Material	CaO	SiO_2	Al_2O_3	Fe_2O_3	MgO	SO_3	R_2O
Low heat cement	58.7	22.8	4.3	4.3	4.2	3.0	0.3
Moderate heat cement	47.9	25.1	11.3	2.4	5.5	3.0	1.3
Fly ash	3.2	52.4	24.0	9.4	1.1	0.4	0.9

2.4. Experimental Procedure

Two batches of temperature stress tests were designed for LHC and MHC separately. Four TSTM were employed in each experimental set using a multi-TSTM system [29]. The restraint control methods used by each TSTM are drawn in Figure 4a for LHC and Figure 4b for MHC.

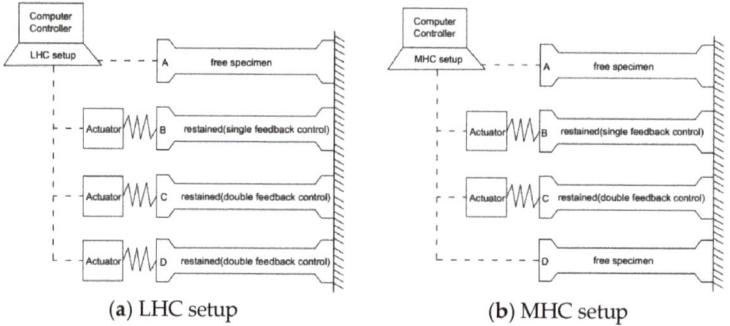

(a) LHC setup   (b) MHC setup

Figure 4. TSTM setups for the testing of low-heat cement concrete (LHC) and moderate-heat cement concrete (MHC) specimens.

The specimens tested by the TSTM A for LHC and tested by TSTM A and TSTM D for MHC were all set as free ones. The specimens on other TSTMs were set as restrained. Note that the restrained specimens were controlled by one of two methods, single or double feedback. As shown in Figure 4a,b. TSTM B for LHC and MHC were controlled by the single feedback method. Both TSTM C and TSTM D for LHC were controlled with double feedback method to verify the reproducibility of the double feedback method. The MHC-A and MHC-D TSTMs were set with free restraints to verify that the free deformation measured by the TSTMs was representative.

To simulate a realistic temperature history such as that found at the construction site of a dam, typical temperature history curves measured at the subject site were used as shown in Figure 5. To ensure that the placement temperature was between 14 °C and 16 °C, which was required by the temperature control strategy of the dam construction, all the materials were precooled in an artificial climate laboratory atmosphere of 0 °C for 24 h before mixed. The temperature was reduced at a rate of 0.5 °C/h after the concrete specimens had been cured for seven days.

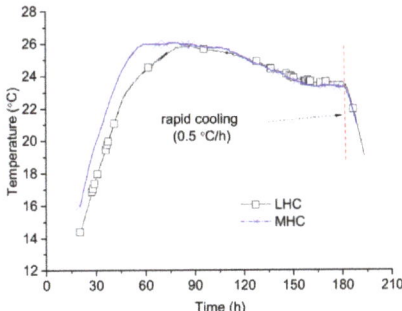

Figure 5. The controlled temperature curves of LHC and MHC measured at a dam construction site.

The restrained test should be initiated as early as possible in order to determine the effects of the early age properties; however, premature failure can occur if the concrete is not of sufficient strength. A variety of methods were studied to determine an appropriate starting time (t_0 or "zero time") with respect to aspects of the restrained stress increment [27,28], temperature rate [34], autogenous deformation rate [33,35], and the earliest possible time [3]. The starting time finally selected was approximately equal to the final setting time [36]. In this work, the initial setting time was 14 h and the final setting time was 20 h, so the starting time was chosen as 20 h after the specimens were cured, and at that time the deformation was established as zero. The side formworks were removed 2 h before the starting time to minimize the temperature gradient between the concrete and the surrounding environment.

The concrete was poured into the TSTM and covered by plastic sheets as shown in Figure 1b. After the side formworks were removed, the side surfaces of LHC and MHC specimens were still covered by plastic sheets as shown in Figure 1b. The top surface of the LHC specimens was then sealed with self-sticking aluminum foil to prevent drying, while the top surface of the MHC specimens remained only covered by plastic sheeting in order to conduct a preliminary investigation of drying creep. The detailed experimental procedure of concrete preparation and TSTM protocol can be found in reference [29]. The measured deformation distance of the specimen was 1000 mm, even though the actual length of the specimen was 2000 mm, and a uniform distribution of the restraint stress was obtained between the 1000 mm, of which the schematic of TSTM is drawn in Figure 1a. The deformation sensor has a 0.1 μm resolution and 0.2 μm reproduction accuracy. The concrete temperature was measured by three temperature sensors inserted into the specimen, in the locations shown position can be seen in Figure 1a,b.

3. Results and Discussion

3.1. Restrained Stress

The restrained stress histories of the LHC and MHC specimens are shown in Figure 6. The measured stress difference between LHC-C and LHC-D is insignificant. Both of these TSTMs were controlled by the double feedback method, which indicates that the stresses measured by different TSTMs are reproducible. During the compensation cycles, the stress of specimen LHC-B was varied while that of LHC-C and LHC-D was held constant. Specimens MHC-B and MHC-C manifested the same pattern as the LHC specimen. The results indicate that the restrained stress is variable between the compensation cycles under the single feedback control method.

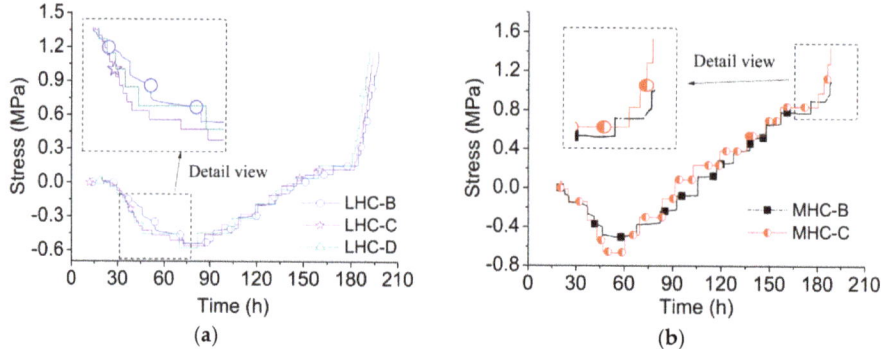

Figure 6. Restrained stress history measured by TSTMs. (**a**) The stress history of LHC; (**b**) The stress history of MHC.

Both the restrained stress and strain are variable between compensation cycles, so the temperature stress test is like a kind of creep relaxation combined experiment. The developed double feedback control method holds the load constant, which results in a stepwise creep experimental process. In the temperature increase phase, the compressive stress measured by the single feedback method increased slower than that measured by the double feedback method, as shown in Figure 6a,b During the temperature decrease phase, the tensile stress measured by the single feedback method was smaller than that measured by the double feedback method. Figure 6b shows that the tensile stress of MHC-B was only 1.09 MPa at failure, while specimen MHC-C exhibited a tensile stress of 1.45 MPa at failure. The inset detail view in Figure 6a shows that the stress in LHC-B lags behind that in LHC-C, and the inset detail in Figure 6b shows the same trend for the MHC specimens.

3.2. Free and Restrained Deformation

The cumulative deformation results of the LHC and MHC specimen tests are shown in Figure 7a,b, in which positive deformations indicate expansion and negative deformations indicate shrinkage. The value of the free deformation was observed to increase with the temperature and vice versa. In Figure 7a, the maximum free strain is 57.3 $\mu\varepsilon$ when the temperature is 26.60 °C. The maximum compressive cumulative deformations of LHC-C and LHC-D are only about 30 $\mu\varepsilon$ because most of the temperature deformation transforms into creep. When the temperature cools from 26.60 °C to 19.50 °C, the free deformation decreases from 57.3 $\mu\varepsilon$ to 2.2 $\mu\varepsilon$ and the restrained deformation decreases from 30 $\mu\varepsilon$ to -9.4 $\mu\varepsilon$.

The restrained deformation of LHC-B is smaller than that of LHC-C and LHC-D, corroborating the understanding that the stress determined by the single feedback method lags behind that determined by the double feedback method. An additional restrained elastic deformation is produced in the single feedback method, so the restrained deformation and stress are smaller than those measured in the double feedback method. In the rapid cooling phase, as the restrained stress increased in the single feedback method, the restrained stress is constant in the double feedback method, and an additional restrained stress would retard the increase in the restrained deformation, as shown in Figure 2b. Hence, the change in restrained deformation controlled by the single feedback method is obviously slower than that controlled by the double feedback method.

The free deformations of MHC-A and MHC-D shown in Figure 7b are almost identical, which indicates that the TSTM system provides good deformation measurement and reproducibility. Though the number of specimens is limited in the temperature stress tests, the results indicate that the proposed TSTM system provides high measurement accuracy. The maximum free strain is 49.8 $\mu\varepsilon$ when the temperature reaches 26.00 °C and decreases to -15.6 $\mu\varepsilon$ as the temperature cools from 26.00 °C to 21.33 °C. The MHC specimens exhibit larger shrinkage than the LHC specimens because the MHC specimens were only sealed by plastic film rather than self-sticking aluminum foil. The free deformation of the MHC specimens contains temperature deformation, autogenous volume deformation, and dry shrinkage deformation. The restrained deformation of MHC-C develops faster than that of MHC-B, which is the same as the LHC, a result of the different feedback control methods.

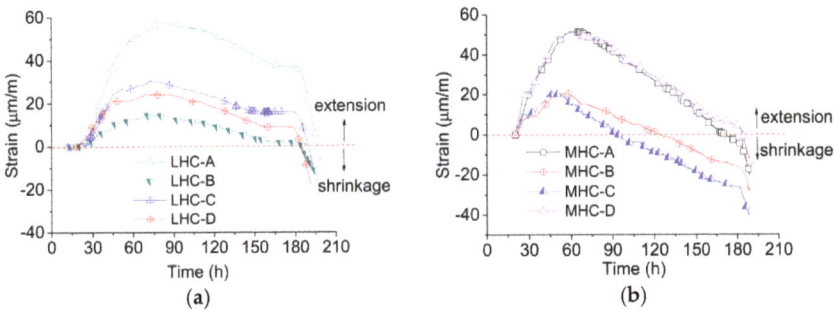

Figure 7. The deformation evolution history from different TSTM tests. (**a**) Deformation results of LHC specimens; (**b**) Deformation results of MHC specimens.

3.3. Restrained Creep

Restrained creep deformations are derived from the free and restrained deformations, with the measurements shown in Figure 8a,b. The creep results of LHC-B, controlled by the single feedback method, are larger than those of LHC-C and LHC-D, controlled by the double feedback method. A restrained elastic stress increment is generated during the compensation cycle, so the creep deformation calculated by Equation (2) includes the restrained elastic deformation, which can be eliminated by Equation (6). The modulus of elasticity (E) is a key factor influencing the results, and the E values of the LHC and MHC were determined at ages of 1, 3, 5, and 7 days by a standard method [37]. In Reference [37], the E values were determined by applying a load of 0.2 MPa/s on the specimen, of which the dimension was $100 \times 100 \times 300$ mm. The E value was calculated from the variation in stress amplitude from 0.5 MPa to 40% of the specimen compressive strength at the loading age. The values of E provided in Table 4 were calculated based on the average values from three $100 \times 100 \times 300$ mm specimen tests at the given testing age.

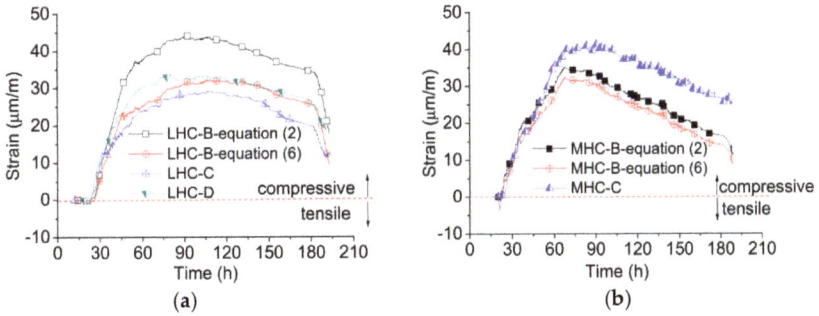

Figure 8. The restrained creep from different TSTM tests. Compressive creep is indicated as a positive value. (**a**) Creep results of LHC; (**b**) Creep results of MHC.

Table 4. The modulus of elasticity (E) of the low-heat cement concrete (LHC) and moderate-heat cement concrete (MHC) at different ages (GPa).

Concrete		1 Day	3 Days	5 Days	7 Days
LHC	Average value	10.57	14.27	16.83	21.73
	Measured values	11.60, 11.70, 8.40	12.20, 14.40, 16.20	15.30, 16.70, 18.00	19.00, 22.50, 22.00
MHC	Average value	13.90	23.27	24.27	26.67
	Measured values	13.70, 13.80, 14.20	23.70, 23.00, 23.10	24.10, 23.50, 25.22	24.90, 29.50, 25.60

The loading stress amplitude affects the determined value of E, and the degree of influence of the stress/strength ratio on the E value is different at different ages [38]. The stress level and stress increments in the specimens during the temperature stress tests were quite small, so the initial tangent E-modulus appeared to be more relevant for use in the stress calculations [33] than those determined from the standard E-modulus [37]. The E value determined using the lower stress amplitude is larger than that using the high-stress amplitude. However, the E values calculated by stress and strain increments from the TSTM yield larger scatter because the increments are so small. The "active method" was then applied [39] by cyclically applying a compressive load corresponding to approximately 10% of the concrete compressive strength. Using this method, the value of E looks very similar to the E values calculated using the stress and strain increments and is approximately 130% of the E values determined by the standard method [37], which are also in accordance with existing research results [40]. Therefore, the E values used for calculation in Equation (6) were set to 130% of the values shown in Table 4.

After considering the restrained elastic deformation, the maximum compressive creep of LHC-B can be observed to decrease from 45 με to 30 με as shown in Figure 8a and is almost equal to that of LHC-C and LHC-D. The elimination of elastic strain is clearly necessary for LHC: The creep results from the single feedback method calculated by Equation (6) match well with those of the double feedback method for LHC. The restrained creep of MHC-B is smaller than that of MHC-C, especially after 60 h, as shown in Figure 8b. Because the restrained deformation of MHC-C varies faster than that of MHC-B in Figure 7b, the creep value of MHC-B, determined by Equation (2), is smaller than that of MHC-C. This difference is caused by material dispersion and is not contrary to the behavior of LHC. Because the E values of MHC are much larger than those of LHC, the restrained elastic deformation of MHC determined by Equation (5) is smaller than that of LHC. The restrained creep results from MHC-B determined by Equations (2) and (6) show little difference. These results indicate that effects of restrained elastic deformation on restrained creep are more obvious for concrete with a lower E. This influence can, thus, be safely neglected for high strength concrete as has been done by some researchers [2,41]. The double feedback method, proposed for determining the restrained creep at early ages, is generally confirmed to be accurate by the creep results of the single feedback method after considering the effects of restrained elastic deformation.

The restrained creep can influence the magnitude of the restrained stress. The restrained stress can be measured by the load cell of the TSTM directly, and the theoretical elastic stress can then be derived by Equation (7).

$$\sigma_{fe}(t_n) = \sum_{i=1}^{n} \Delta\varepsilon_{free}(t_i) \cdot E(t_i) \qquad (7)$$

where $\sigma_{fe}(t_n)$ is the theoretical elastic stress at a time t_n corresponding to the free deformation, and $E(t_i)$ is 130% of the values given in Table 4, according to references [39,40]. The maximum compressive stress in the LHC in Figure 6a is only between 0.54 and 0.58 MPa. However, when the temperature of the concrete increases to its maximum value of 26.60 °C, the maximum free deformation is 57.3 με, corresponding to a theoretical elastic stress of approximately 0.95 MPa. The relaxation is about 40–45% at the maximum temperature. The maximum compressive restrained creep is 25–30 με, and the ratio between the creep and free deformation of LHC is about 43.6–52.4%, which is very close to the relaxation ratio. Similarly, the ratio between the creep and free deformation of MHC is about 50–60%, and the stress relaxation ratio is also about 50–60%. The results indicate that, no matter the material, the restrained creep is 40–60% of the free deformation in the restrained temperature stress test, and the creep-to-free deformation ratio is approximately equal to the stress relaxation degree under the given temperature history. This 40–60% creep-to-free deformation ratio has also been observed in high performance concretes [2,3,41,42] whether or not the temperature histories were constant or variable. Thus, the ratio between creep and free deformation can be used as a measure of stress relaxation [7], and early-age creep could relax 40–60% of the restrained stress at early age.

A TSTM can determine the restrained creep as influenced by the combined factors of temperature and creep recovery from a very early age, which can hardly be obtained using only a test with the conventional creep testing method of applying a constant load [41]. The classical linear viscoelasticity creep theory [43–45] can also predict restrained creep, however, the viscoelasticity creep model should consider the combined factors such as variable temperature, alternating tension and compression stress, and the early age effects. A significant amount of testing is required to obtain a creep model that can be used to calculate the restrained creep exactly. Some researchers [46,47] have predicted the restrained stress using a creep model derived from the conventional testing method. However, Kovler [41] found that specific creep values obtained in the plateau from a restrained shrinkage test were smaller than those obtained from a conventional test. Two reasons may account for the difficulty of combining the restrained creep as determined by TSTM with that determined by the conventional method. One reason is that the magnitude of self-induced restrained stress is small. The restrained stress at very early age is near zero, yet the restrained creep is considerable. Therefore, the specific creep as determined by dividing creep strain by restrained stress would result in a large discrepancy. Another reason is that the restrained creep is influenced by many factors, so the classical method, which obtains results by applying only a constant load, is insufficient.

In future work, a creep model considering variable temperature, alternating tension–compression stress, and early age effects will be studied by directly measuring creep and free strains. Furthermore, the restrained creep can be accurately predicted by combining the creep model.

4. Conclusions

A double feedback control method using a TSTM was developed based on a multi-TSTM system to measure the early-age restrained creep of LHC and MHC specimens. The conclusions are as follows:

(1) The TSTM is a very useful instrument for determining the early-age restrained creep of dam concrete under the combined effects of varying temperature, creep recovery, tension and compression stress, and early age. The double feedback control method for a TSTM can hold the stress constant during the compensation cycle, eliminating the restrained elastic deformation.

(2) The restrained creep measured by the conventional single feedback TSTM control method neglects the effects of restrained elastic deformation, and as a result, the measured creep is larger than the actual value. A new creep calculation equation for single feedback method was, accordingly, derived. The results calculated with the proposed equation show positive agreement with the results of the more accurate double feedback method.

(3) The difference in restrained creep between single and double feedback methods is significant in low elastic modulus concrete and can be neglected in high elastic modulus concrete.

In summary, the double feedback control method is recommended for determining the restrained creep of concrete using a TSTM. The value of early-age restrained creep is significant and should not be neglected. The ratio between creep and free deformation at an early age may be in the range of 40–60% for LHC, MHC, and high-performance concretes.

Author Contributions: H.Z., Q.L. and Y.H. developed the TSTM system. H.Z., Q.L., Y.H. and R.M. conceived and designed the experiments. H.Z. performed the experiments, analyzed the data and wrote the initial draft of the manuscript; Q.L. and Y.H. reviewed and contributed to the final manuscript.

Funding: National Natural Science Foundation of China (No. 51579134 and No. 51339003), and the Research Fund of Tsinghua University (Grant No. 20161080079).

Acknowledgments: This work was supported by the National Natural Science Foundation of China (No. 51579134 and No. 51339003), and the Research Fund of Tsinghua University (Grant No. 20161080079). The costs to publish in open access have been covered by funding.

Conflicts of Interest: The authors declare no conflict of interest.

References

1. Bofang, Z. *Thermal Stresses and Temperature Control of Mass Concrete*, 1st ed.; Butterworth-Heinemann: Waltham, MA, USA, 2013; pp. 1–10. ISBN 978-0-12-407723-2.
2. Tao, Z.; Weizu, Q. Tensile creep due to restraining stresses in high-strength concrete at early ages. *Cem. Concr. Res.* **2006**, *36*, 584–591. [CrossRef]
3. Altoubat, S.A.; Lange, D.A. Creep, shrinkage, and cracking of restrained concrete at early age. *ACI Mater. J.* **2001**, *98*, 323–331.
4. Shen, D.; Jiang, J.; Wang, W.; Shen, J.; Jiang, G. Tensile creep and cracking resistance of concrete with different water-to-cement ratios at early age. *Constr. Build. Mater.* **2017**, *146*, 410–418. [CrossRef]
5. Li, K.; Ju, Y.; Han, J.; Zhou, C. Early-age stress analysis of a concrete diaphragm wall through tensile creep modeling. *Mater. Struct.* **2009**, *42*, 923–935. [CrossRef]
6. Li, Q.; Liang, G.; Hu, Y.; Zuo, Z. Numerical analysis on temperature rise of a concrete arch dam after sealing based on measured data. *Math. Probl. Eng.* **2014**, *2014*, 602818. [CrossRef]
7. Atrushi, D.S. Tensile and Compressive Creep of Early Age Concrete: Testing and Modelling. Ph.D. Thesis, the Norwegian University of Science and Technology, Trondheim, Norway, 2003.
8. Altoubat, S.A.; Lange, D.A. Tensile basic creep: Measurements and behavior at early age. *ACI Mater. J.* **2001**, *98*, 386–393.
9. Pane, I.; Hansen, W. Investigation on key properties controlling early-age stress development of blended cement concrete. *Cem. Concr. Res.* **2008**, *38*, 1325–1335. [CrossRef]
10. Hauggaard, A.B.; Damkilde, L.; Hansen, P.F. Transitional thermal creep of early age concrete. *J. Eng. Mech.* **1999**, *125*, 458–465. [CrossRef]
11. Sabeur, H.; Meftah, F. Dehydration creep of concrete at high temperatures. *Mater. Struct.* **2008**, *41*, 17–30. [CrossRef]
12. Bazant, Z.P.; Baweja, S. Creep and shrinkage prediction model for analysis and design of concrete structures: Model B3. *ACI Spec. Publ.* **2000**, *194*, 1–84.
13. Bazant, Z.P.; Cusatis, G.; Cedolin, L. Temperature effect on concrete creep modeled by microprestress-solidification theory. *J. Eng. Mech.* **2004**, *130*, 691–699. [CrossRef]
14. Ladaoui, W.; Vidal, T.; Sellier, A.; Bourbon, X. Effect of a temperature change from 20 to 50 °C on the basic creep of HPC and HPFRC. *Mater. Struct.* **2011**, *44*, 1629–1639. [CrossRef]
15. Vidal, T.; Sellier, A.; Ladaoui, W.; Bourbon, X. Effect of temperature on the basic creep of high-performance concretes heated between 20 and 80 °C. *J. Mater. Civ. Eng.* **2014**, *27*, B4014002. [CrossRef]
16. Schutter, G.D.; Yuan, Y.; Liu, X.; Jiang, W. Degree of hydration-based creep modeling of concrete with blended binders: From concept to real applications. *J. Sustain. Cem. Based Mater.* **2015**, *4*, 1–14. [CrossRef]
17. Jiang, W.; Schutter, G.D.; Yuan, Y. Degree of hydration based prediction of early age basic creep and creep recovery of blended concrete. *Cem. Concr. Compos.* **2014**, *48*, 83–90. [CrossRef]
18. Briffaut, M.; Benboudjema, F.; Torrenti, J.; Nahas, G. Concrete early age basic creep: Experiments and test of rheological modelling approaches. *Constr. Build. Mater.* **2012**, *36* (Suppl. C), 373–380. [CrossRef]
19. Hermerschmidt, W.; Budelmann, H. Creep of early age concrete under variable stress. In Proceedings of the 10th International Conference on Mechanics and Physics of Creep, Shrinkage, and Durability of Concrete and Concrete Structures, Vienna, Austria, 21–23 September 2015; American Society of Civil Engineers (ASCE): Vienna, Austria, 2015; pp. 929–937. [CrossRef]
20. Delsaute, B.; Boulay, C.; Staquet, S. Creep testing of concrete since setting time by means of permanent and repeated minute-long loadings. *Cem. Concr. Compos.* **2016**, *73*, 75–88. [CrossRef]
21. Delsaute, B.; Torrenti, J.; Staquet, S. Modeling basic creep of concrete since setting time. *Cem. Concr. Compos.* **2017**, *83*, 239–250. [CrossRef]
22. Østergaard, L.; Lange, D.A.; Altoubat, S.A.; Stang, H. Tensile basic creep of early-age concrete under constant load. *Cem. Concr. Res.* **2001**, *31*, 1895–1899. [CrossRef]

23. Wei, Y.; Guo, W.; Liang, S. Microprestress-solidification theory-based tensile creep modeling of early-age concrete: Considering temperature and relative humidity effects. *Constr. Build. Mater.* **2016**, *127*, 618–626. [CrossRef]
24. Springenschmid, R.; Breitenbücher, R. Are low heat cements the most favourable cements for the prevention of cracks due to heat of hydration? *Concr. Precast. Plant Technol.* **1986**, *52*, 704–711.
25. Kovler, K. Testing system for determining the mechanical behaviour of early age concrete under restrained and free uniaxial shrinkage. *Mater. Struct.* **1994**, *27*, 324–330. [CrossRef]
26. Klausen, A.E.; Kanstad, T.; Bjøntegaard, Ø.; Kollegger, J.; Hellmich, C.; Pichler, B. Updated Temperature-Stress Testing Machine (TSTM): Introductory Tests, Calculations, Verification, and Investigation of Variable Fly Ash Content. In Proceedings of the 10th International Conference on Mechanics and Physics of Creep, Shrinkage, and Durability of Concrete and Concrete Structures, Vienna, Austria, 21–23 September 2015; American Society of Civil Engineers (ASCE): Vienna, Austria, 2015; pp. 724–732. [CrossRef]
27. Staquet, S.; Delsaute, B.; Darquennes, A.; Espion, B. Design of a revisited TSTM system for testing concrete since setting time under free and restraint conditions. In Proceedings of the Concrack3—RILEM-JCI International Workshop on Crack Control of Mass Concrete and Related Issues Concerning Early-Age of Concrete Structures, Paris, France, 15–16 March 2012; Concrack3—RILEM-JCI: Paris, France, 2012; pp. 99–110.
28. Charron, J.P.; Bissonnette, B.; Marchand, J.; Pigeon, M. Test device for studying the early-age stresses and strains in concrete. *ACI Spec. Publ.* **2004**, *220*, 113–124. [CrossRef]
29. Zhu, H.; Li, Q.; Hu, Y. Self-developed testing system for determining the temperature behavior of concrete. *Materials* **2017**, *10*, 419. [CrossRef] [PubMed]
30. Wei, Y.; Liang, S.; Guo, W. Decoupling of autogenous shrinkage and tensile creep strain in high strength concrete at early ages. *Exp. Mech.* **2017**, *57*, 475–485. [CrossRef]
31. Klausen, A.E.; Kanstad, T.; Bjøntegaard, Ø.; Sellevold, E. Comparison of tensile and compressive creep of fly ash concretes in the hardening phase. *Cem. Concr. Res.* **2017**, *95*, 188–194. [CrossRef]
32. Wei, Y.; Hansen, W. Tensile creep behavior of concrete subject to constant restraint at very early ages. *J. Mater. Civ. Eng.* **2013**, *25*, 1277–1284. [CrossRef]
33. Bjøntegaard, Ø.; Sellevold, E.J. The temperature-stress testing machine (TSTM): Capabilities and limitations. In Proceedings of the First International RILEM Symposium on Advances in Concrete Through Science and Engineering, Evanston, IL, USA, 21–24 March 2004; Weiss, J., Kovler, K., Marchand, J., Eds.; RILEM Publications SARL: Evanston, IL, USA, 2004.
34. Cusson, D.; Hoogeveen, T. An experimental approach for the analysis of early-age behaviour of high-performance concrete structures under restrained shrinkage. *Cem. Concr. Res* **2007**, *37*, 200–209. [CrossRef]
35. Pigeon, M.; Toma, G.; Delagrave, A.; Bissonnette, B.; Marchand, J.; Prince, J.C. Equipment for the analysis of the behaviour of concrete under restrained shrinkage at early ages. *Mag. Concr. Res.* **2000**, *52*, 297–302. [CrossRef]
36. Kanstad, T.; Hammer, T.A.; Bjøntegaard, Ø.; Sellevold, E.J. Mechanical properties of young concrete: Part II: Determination of model parameters and test program proposals. *Mater. Struct.* **2003**, *36*, 226–230.
37. The Ministry of Water Resources of the People's Republic of China. *Test Code for Hydraulic Concrete*; SL 352-2006; China Water Power Press: Beijing, China, 2006.
38. Delsaute, B.; Boulay, C.; Granja, J.; Carette, J.; Azenha, M.; Dumoulin, C.; Karaiskos, G.; Deraemaeker, A.; Staquet, S. Testing concrete E—Modulus at very early ages through several techniques: An inter—Laboratory comparison. *Strain* **2016**, *52*, 91–109. [CrossRef]
39. Zhu, H.; Hu, Y.; Li, Q.; Zhang, M. Determination of concrete elastic modulus in early age for temperature stress testing under the effect of restraint. In Proceedings of the 2nd International RILEM/COST Conference on Early Age Cracking and Serviceability in Cement-Based Materials and Structures—EAC2, Brussels, Belgium, 12–14 September 2017; ULB and VUB: Brussels, Belgium, 2017.
40. Mehta, P.K.; Monteiro, P.J.M. *Concrete: Microstructure, Properties, and Materials*, 4th ed.; McGraw-Hill Publishing: New York, NY, USA, 2006; pp. 64–70. ISBN 1281080799.
41. Kovler, K.; Igarashi, S.; Bentur, A. Tensile creep behavior of high strength concretes at early ages. *Mater. Struct.* **1999**, *32*, 383–387.

42. Shen, D.J.; Jiang, J.L.; Shen, J.X.; Yao, P.P.; Jiang, G.Q. Influence of prewetted lightweight aggregates on the behavior and cracking potential of internally cured concrete at an early age. *Constr. Build. Mater.* **2015**, *99*, 260–271. [CrossRef]
43. Christensen, R. *Theory of Viscoelasticity: An Introduction*; Elsevier: Dordrecht, the Netherlands, 2012; ISBN 0323161820.
44. Gilbert, R.I. *Time Effects in Concrete Structures*; Elsevier: New York, NY, USA, 1988; Volume 23, ISBN 0167-6288.
45. Neville, A.M.; Dilger, W.H.; Brooks, J.J. *Creep of Plain and Structural Concrete*; Construction Press: London, UK, 1983; ISBN 9780860958345.
46. Pane, I.; Hansen, W. Predictions and verifications of early-age stress development in hydrating blended cement concrete. *Cem. Concr. Res.* **2008**, *38*, 1315–1324. [CrossRef]
47. Wei, Y.; Liang, S.; Guo, W.; Hansen, W. Stress prediction in very early-age concrete subject to restraint under varying temperature histories. *Cem. Concr. Compos.* **2017**, *83*, 45–56. [CrossRef]

© 2018 by the authors. Licensee MDPI, Basel, Switzerland. This article is an open access article distributed under the terms and conditions of the Creative Commons Attribution (CC BY) license (http://creativecommons.org/licenses/by/4.0/).

MDPI
St. Alban-Anlage 66
4052 Basel
Switzerland
Tel. +41 61 683 77 34
Fax +41 61 302 89 18
www.mdpi.com

Materials Editorial Office
E-mail: materials@mdpi.com
www.mdpi.com/journal/materials

www.ingramcontent.com/pod-product-compliance
Lightning Source LLC
LaVergne TN
LVHW070645100526
838202LV00013B/882

MDPI
St. Alban-Anlage 66
4052 Basel
Switzerland
Tel. +41 61 683 77 34
Fax +41 61 302 89 18
www.mdpi.com

Materials Editorial Office
E-mail: materials@mdpi.com
www.mdpi.com/journal/materials